Structural and Chemical Features of Chalcogenides

Edited by Suresh Sagadevan

Published in London, United Kingdom

Structural and Chemical Features of Chalcogenides
http://dx.doi.org/10.5772/intechopen.111104
Edited by Suresh Sagadevan

Contributors
Adeyeba Adewale Oluwatobi, Arunkumar Priya, Edgar Mosquera, Emmanuel Ifeanyi Ugwu, Geetha Kaliyan, Ibrahim Abdullahi Danasabe, Manivel Rajan, Mohamend Jaafar, Nour El Houda Safi, Rafiu Adewale Busari, Raja Arumugam, Rajesh Paulraj, Ramasamy Perumalsamy, Ramesh Sivasamy, Selvam Kaliyamoorthy, Sivasubramani Vediyappan, Siva Vadivel, Sunday Ikpughul Lyua, Suresh Sagadevan, Usman Rilwan

First published in London, United Kingdom, 2024 by IntechOpen
IntechOpen is the global imprint of INTECHOPEN LIMITED, registered in England and Wales, registration number: 11086078, 167-169 Great Portland Street, London, W1W 5PF, United Kingdom

British Library Cataloguing-in-Publication Data
A catalogue record for this book is available from the British Library

Additional hard and PDF copies can be obtained from orders@intechopen.com

Structural and Chemical Features of Chalcogenides
Edited by Suresh Sagadevan
p. cm.
Print ISBN 978-0-85466-185-5
Online ISBN 978-0-85466-184-8
eBook (PDF) ISBN 978-0-85466-186-2

For EU product safety concerns:
IN TECH d.o.o., Prolaz Marije Krucifikse Kozulić 3, 51000 Rijeka, Croatia,
info@intechopen.com or visit our website at intechopen.com.

Meet the editor

Dr. Suresh Sagadevan is an associate professor at the Nanotechnology and Catalysis Research Centre, University of Malaya. He has published more than 450 research papers in the ISI's top-tier journals and Scopus. He has authored 15 international book series and 50 book chapters. He is an editor, guest editor, and editorial board member of many reputed ISI journals. He is a member of many professional bodies at the national and international level. He has been a recognized reviewer for many reputed journals. He is also working in various fields such as nanofabrication, functional materials, graphene, polymeric nanocomposite, glass materials, thin films, bio-inspired materials, drug delivery, tissue engineering, cell culture, supercapacitor, optoelectronics, photocatalytic, green chemistry, and biosensor applications.

Contents

Preface

The chalcogenides are one of the fascinating classes of materials composed of elements from group 16 of the periodic table (oxygen, sulfur, selenium, tellurium, and polonium), creating interest among scientists over the decades. Their unique structural and chemical properties have led to a remarkable diversity of applications, with profound impacts in fields ranging from electronics and photonics to energy conversion and catalysis.

This book, *Structural and Chemical Features of Chalcogenides*, delves into the captivating world of these materials. We aimed to provide a comprehensive overview of the intricate associations between the structure and chemistry of chalcogenides.

This volume is intended for researchers, students, and professionals working in materials science, chemistry, physics, and other related disciplines.

For the novice, the introductory chapters provide a foundational understanding of chalcogenide chemistry and crystal structures. More advanced sections investigate deeper into complex phenomena, such as bonding, defect chemistry, and the influence of composition on material properties.

Throughout the book, we highlight the interplay between structural features and chemical behavior. We explore the influence of various factors, such as bond lengths, coordination geometries, and the presence of defects, properties such as electrical conductivity, optical response, and thermal stability.

I assure readers that this book will serve as a valuable resource for anyone seeking to understand the captivating world of chalcogenides. The main motive is not only to provide knowledge on the fascinating world of chalcogenides but also to spark curiosity and inspire further exploration of these remarkable materials.

Dr. Suresh Sagadevan, Ph.D., CChem, FRSC
Nanotechnology and Catalysis Research Centre (NANOCAT),
University of Malaya,
Kuala Lumpur, Malaysia

Chapter 1

Assessment of the Annealing Effect on Varied Ligand Deposited Lead Sulphide Thin Film Grown by Chemical Bath Deposition Technique

Emmanuel Ifeanyi Ugwu, Adeyeba Adewale Oluwatobi, Rafiu Adewale Busari, Sunday Ikpughul Lyua, Ibrahim Abdullahi Danasabe, Mohamend Jaafar and Usman Rilwan

Abstract

The study of the influence varied concentrations of ligand and annealing of chemically bath deposited PbS thin film has been carried out the influence of ligand on the optical, solid state, morphological and Electrical properties of the deposited thin films has been investigated. The absorbance, transmittance from the UV-VIS spectrophotometer and other parameters such as absorption coefficient, reflectance, energy band gap, optical conductivity indicate that there is a effect of ligand variation on the properties and the absorption coefficient of the films was found to increase as the concentration of ligand as they annealed to the same temperature and in all case they decreases towards longer wavelength region and the energy band gap was observed to increase from 1.61 to 3.71 eV as the concentration of ligand increases. The surface morphology of the SEM micrographs result shows that the deposited thin films are of polycrystalline and the particle grains are evenly distributed across the substrate's surface indicates that the grain sizes of the films are dependent on the molar concentration of ligand increases while the XRD showed where the preferential plane with highest diffraction intensity is located.

Keywords: lead sulphide, thin-film, varied ligand, assessment chemical bath deposition, annealing

1. Introduction

It has been recently observed that much oxide based and chalcogenide based thin films form the major types of compounds/thin films that have attracted the interest of many researchers due to their anticipated amenability in modern thin film technological applications in this modern period. This observation is based on their unique characteristics which is the root of their application [1]. And apart from these their multipurpose applications, they have characteristic of crystallization with the perovkite structure coupled with their unique behavior in gas-solid interface, heterogeneous catalysis and also their susceptibility in partial substitution in A and B position [2]. They usually have an exceptional capacity of glass forming, a high refractive index, and photonic energy lower than the standards, high photosensitivity, and excellent transmission over the infrared region. These characteristics make them ideal for manufacturing various civil, medical and military applications such as infrared detectors, infrared lens, planar optics, photonics integrated circuit lasers etc. [3]. Coupled with the fact of their abundance on the earth and environmental friendliness, excellent structural, optical and electrical properties, chalcogenide materials are considered to possess excellent absorption coefficient that makes it suitable for use in absorption of photons from sun radiation [4]; they can be synthesized easily with given extensive range of their amenability of their chemical and physical properties.

However, our work here is focusing only on binary sulfur chalcogenides based thin film known to be ternary neither quaternary chalcogenide nor is it on oxide base thin film which has also attracted lots of researchers [5].

Binaries chalcogenide constitute the following; CdS, CdSe, Bi_2S_3, PbSe, PbS, As_2S_3, Ag_2S, Cu_2S, Sb_2Se_3, ZnS, CaS, PbTe, MnS. Some of which have been grown using CBD technique [6] and by other deposition techniques such as pulsed laser technique has been employed to deposit PbS thin film [7]; which is an indication that many other techniques have been implemented by many researchers in the growth of this group of thin film showing it's friendly to different growth techniques.

Generally, metal dichalcogenides have the formula ME_2, where M = a transition metal and E = S, Se, Te [8] and the most important members are the sulfides. They are always dark diamagnetic solids, insoluble in all solvents, and exhibit semiconducting properties. Some are superconductors [9]. Several metals, mainly for the early metals (Ti, V, Cr, Mn groups) also form trichalcogenides. These materials are usually described as $M^{4+}(E_2{}^{2-})(E^{2-})$ (where E = S, Se, Te). A well-known example is niobium triselenide. Amorphous MoS_3 is produced by treatment of tetrathiomolybdate with acid [10]. The process involves combination of electropositive elements thin films combined with chalcogens (S, Se and Te) under the control of ligand. For instance, the growth and optical characterization of Antimony Selenide (Sb_2Se_3) and *binaries chalcogenide* such as CdS, CdSe, Bi_2S_3, PbSe, PbS, As_2S_3, Ag_2S, Cu_2S, Sb_2Se_3, ZnS, CaS, PbTe, MnS have been grown using different deposition technique including CBD [6, 11, 12]. It's important to mention here that a more technical approach known as pulsed laser technique has been used to grow PbS in quest to obtain a more refined sample of the film. Also, theoretically and mathematical approach has been used to study and analyze the properties of some of these binary based chalcogenide thin film materials in order to compare the experimental and theoretical analysis [13].

Continuing on the quest for search on optimization of chalcogenide based thin films for applications in harnessing solar energy and for other device applications in this work, we intend to examine the effect of the annealing on varied ligand of chemically bath developed binary PbS thin film and to ascertain its effect on the structural, morphological, electrical and optical properties of thin film.

2. Materials and method

The experimental process and theoretical analysis were carried out as described here in this section.

2.1 Experimental procedure

The chemical bath was prepared by sequential addition of 2.5 ml (0.5 M) of lead acetate trihydrate ($Pb(CH_3COO)_2.3H_2O$), 2.5 ml (2 M) of sodium hydroxide ($NaOH$) to neutralize the acidity of the solution, 3 ml (1 M) of thiourea (CH_4N_2S), while triethanolamine (TEA) as ligand was varied from 0.5 M, 0.8 M to 1.0 M according to **Table 1**. The growth process was carried out at room temperature of (27°C) within a specified period. After preparing the solution as stated above, distilled water was added until the volume of the solution reached 70 ml and the PH of the solution was measured using digital PH meter and kept constant at 11. After which the bath solution was then thoroughly stirred on a magnetic stirrer for 30 min to ensure homogeneity. The cleaned and dried substrate was incited and clamped firmly on a vertical position using retort stand and clip and then lowered into the 100 ml beaker containing the solution inside the beaker with aluminum foil caped on it as a cover for the top of the beaker in order to prevent dust or unwanted particles from entering the solution. The deposition time was kept constant throughout for 12 h for each of the samples labeled T_1, T_2 and T_3 for which the ligand, TEA concentration is 0.5 M, 0.8 M and 1.0 M respectively. At the end of the specified period of the deposition, the three samples were taken out of the bath, rinsed in distilled water and dried in an open furnace at moderate temperature for 3 min to remove residual water content and other possible adsorbed surface impurities. The films were observed to be homogeneous and well adhered to the substrate. **Table 1** gives the details of the concentration of the ligand, (TEA) used for different samples [14, 15].

The optical and solid-state properties to be studied as a thin film characterization includes absorbance, transmittance, reflectance, absorption coefficient, extinction coefficient, refractive index, optical conductivity, dielectric constant, and energy band gap. The study of the solid-state properties of the film would give one an idea of these characteristics which arise as result of the interaction between photon energy and the structure of the thin film or between the energy configuration and other optical constants e.g., refractive index and extinction coefficient of the material [15].

2.2 Theoretical procedure

The optical and solid-state properties were deduced using some of the outlined formulas as given here, for instance, the absorbance A absorbance/transmittance spectra of the films T are obtained directly using spectrophotometer whose relation is given in Eq. (1) [16] although they can be obtained theoretically from the following relations

$$\log_{10}\left(\frac{I}{I_\circ}\right) = \log_{10}\left(\frac{1}{T}\right) = 2 - \log T(\%) \quad (1)$$

Samples	Conc. of ligand (TEA) in (mole)
T_1 (W_1)	0.5
T_2 (W_2)	0.8
T_3 (W_3)	1.0

Table 1.
The Samples showing the arranged sample according to the ligand concentration.

where I represent the incident radiation flux and $I_\circ$ stands for the transmitted flux of the photon energy

$$T = 10^{-A}x100\% \tag{2}$$

Where A is the absorbance.

2.3 Transmittance (T)

Accordingly, transmittance is simply defined as the ratio of the transmitted flux to that of the incident flux I mathematically it is expressed as [17]

$$T = \frac{I}{I_\circ} \tag{3}$$

2.4 Reflectance (R)

Reflectance is determined mathematically from the relation:

$$A + T + R = 1 \tag{4}$$

Where R is the reflectance, A is the absorbance, and T the transmittance. This equation is valid on the account of the principle of energy conservation [16].

2.5 Absorption and extinction coefficient, (α) and (K)

Accordingly, the absorbance and absorption coefficient could be calculated in the fundamental absorption region using Lambert law:

$$\ln\frac{I}{I_\circ} = 2.303A = \alpha d, \tag{5}$$

$$\alpha = \frac{2.303A}{d} \tag{6}$$

$$k = \frac{\alpha\lambda}{4\pi} \tag{7}$$

Where λ is the wavelength of the incident photon.

2.6 Refractive index and optical conductance (n) and (σ)

From the transmittance result analysis, the refractive index n of the sample could be calculated using the following relations;

$$n = \frac{1}{T} + \left(\frac{1}{T-1}\right)^{½} \text{ or } n = \frac{1+\sqrt{R}}{1-\sqrt{R}} \tag{8}$$

where R is defined as [18];

$$R = \frac{(n-1)^2 + k^2}{(n+1)^2 + k^2} \tag{9}$$

and

$$\sigma = \alpha nc\varepsilon_o \tag{10}$$

3. Results/discussion

Figure 1 is the experimental set up used for the deposition process of the film below.

Figures 2 and **3** presents the absorbance and transmittance of thin film and **Figures 4-6** depicts the Tuac plots $(\alpha h\nu)^2(eV/cm)^2$ as a function of photon energy (eV) for sample $W_{1,}$ $(\alpha h\nu)^2(eV/cm)^2$ as a function of photon energy (eV) for sample W_2 and $(\alpha h\nu)^2(eV/cm)^2$ as a function of photon energy (eV) for sample W_3. **Figures 7-9** showcased the refractive index plot as a function of photon energy in eV while the optical conductance for each of the samples is presented in **Figures 10-12** respectively.

Up to date SEM investigation was carried out using SEM as presented in **Figures 13–15** for the entire samples with the Dispersive X-ray Spectroscopy (EDAX) analysis depicted in **Figure 16**.

3.1 Optical and solid-state feature of PbS thin film

From the results the spectral absorbance as seen in **Figure 2** reveals that the film has higher absorbance within the ultraviolet region and then decreases within the optical and near infrared region of EM wave spectrum for the two samples whose ligand concentration is less than 1.0 mole whereas the sample whose ligand is 1.0 mole

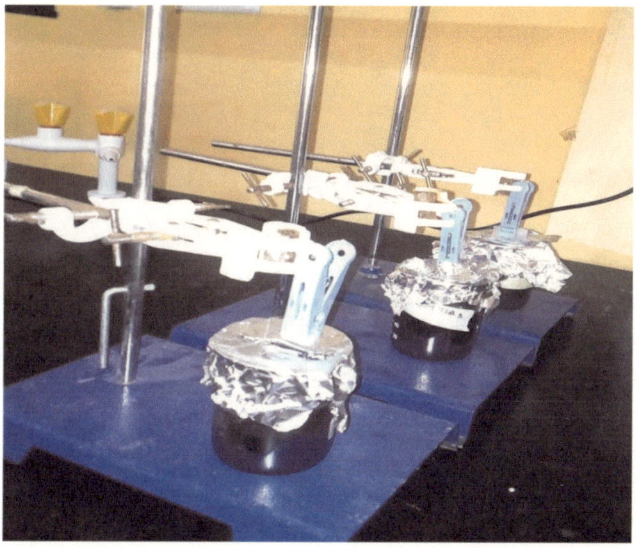

Figure 1.
The three samples clamped with Retort stand during the growth.

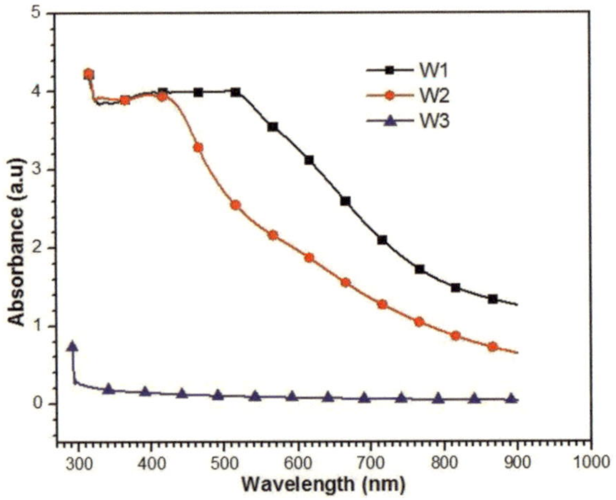

Figure 2.
Absorbance as a function of wavelength for the three samples.

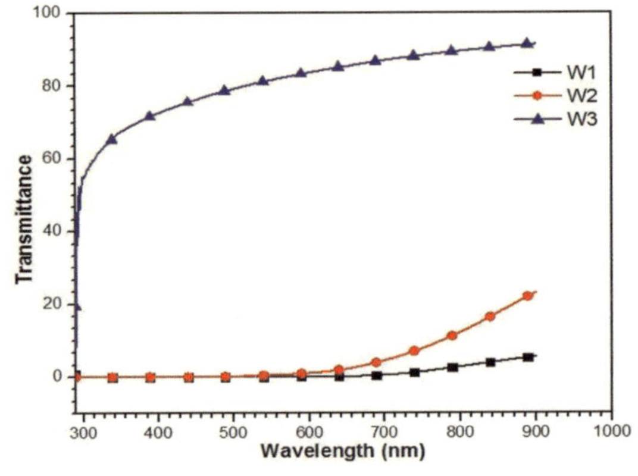

Figure 3.
Transmittance against wave length.

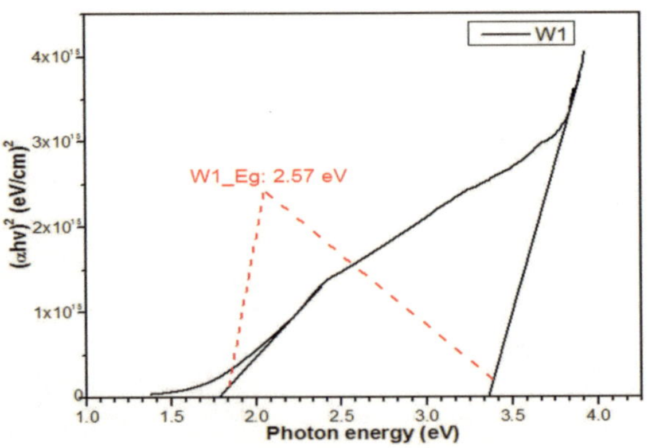

Figure 4.
Plot of $(\alpha h\nu)^2 (eV/cm)^2$ *as a function of photon energy (eV) for sample* W_1.

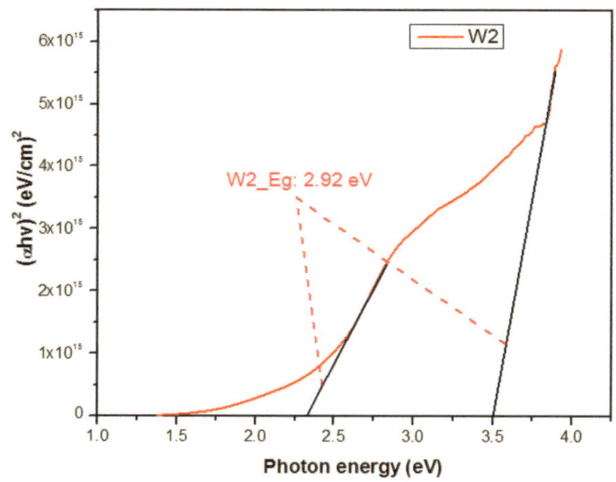

Figure 5.
Plot of $(\alpha h\nu)^2 (eV/cm)^2$ *as a function of photon energy (eV) for sample* W_2.

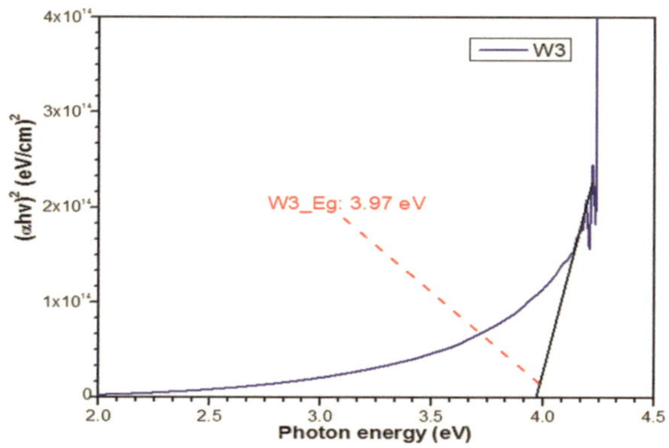

Figure 6.
Plot of $(\alpha h\nu)^2 (eV/cm)^2$ *as a function of photon energy (eV) for sample* W_3.

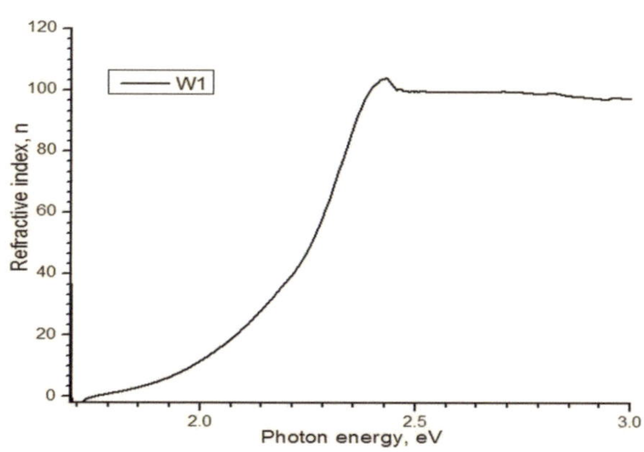

Figure 7.
Refractive index, n as a function photon energy (eV) for W_1.

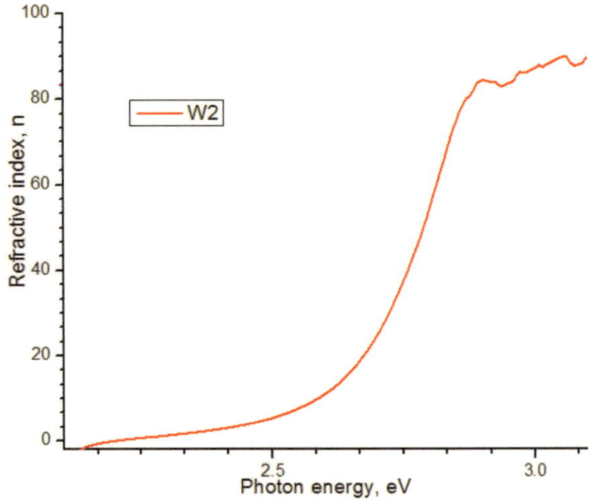

Figure 8.
Refractive index, n as a function photon energy (eV) for W_2.

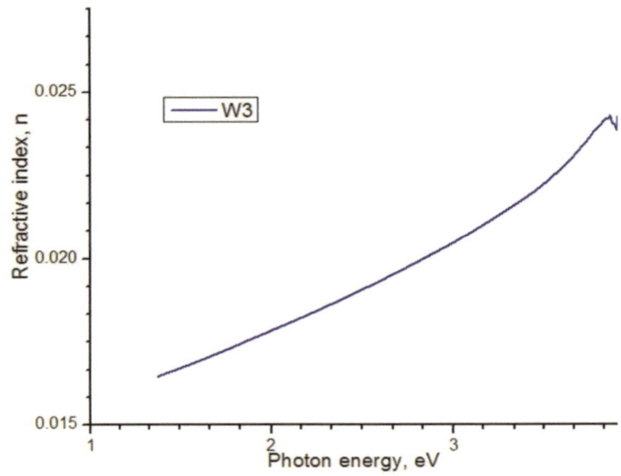

Figure 9.
Refractive index, n as a function photon energy (eV) for W_3.

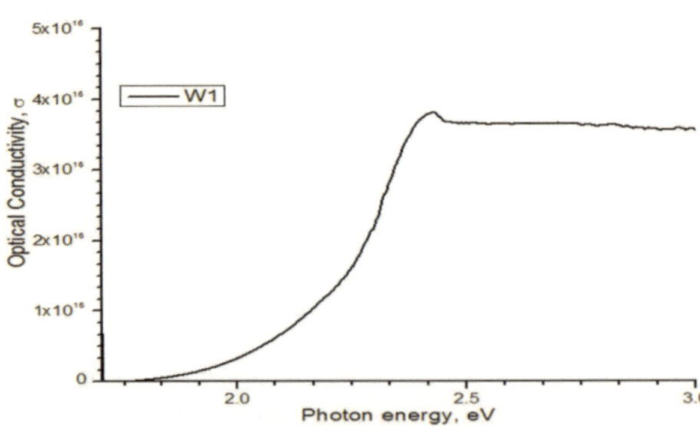

Figure 10.
Optical conductance, α as a function photon energy (eV) for W_1.

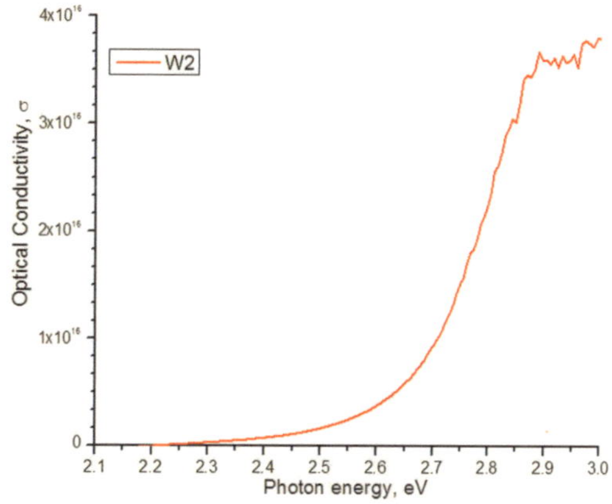

Figure 11.
Optical conductance, α as a function photon energy (eV) for W_2.

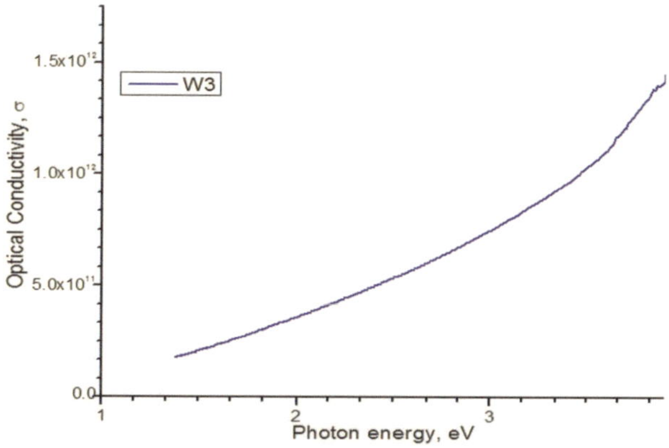

Figure 12.
Optical conductance, α as a function photon energy (eV) for W_3.

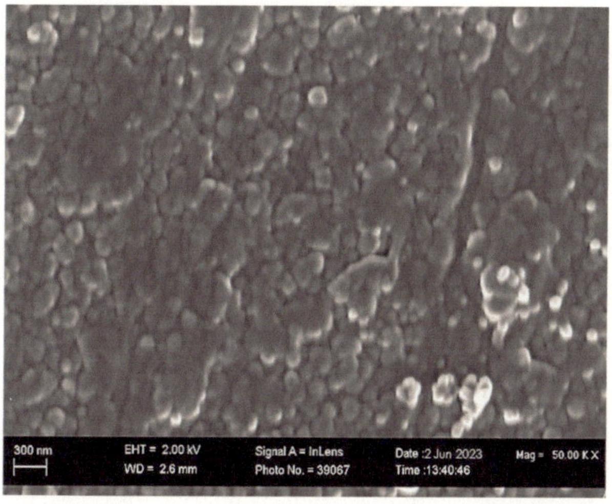

Figure 13.
SEM showing the annealed morphological structure of W_1.

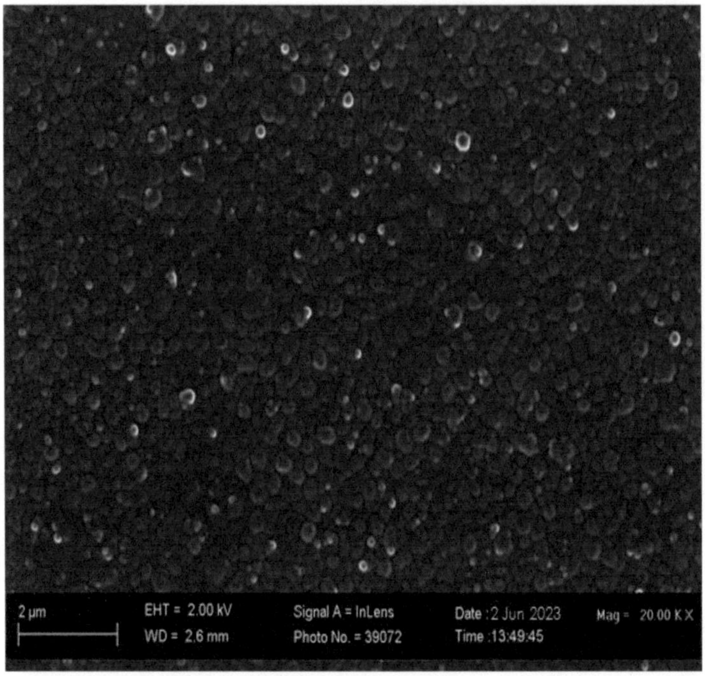

Figure 14.
SEM showing the annealed morphological structure of W_2.

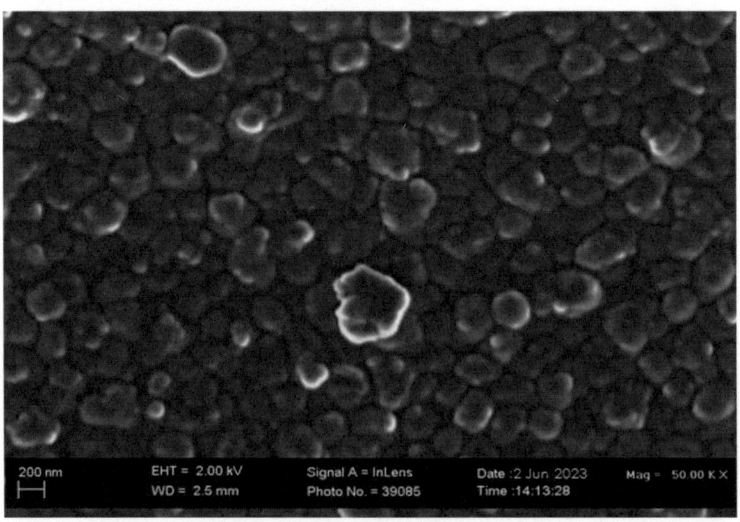

Figure 15.
SEM showing the annealed morphological structure of W_3.

low absorbance for all the spectra. On the other hand, the transmittance appears to be high within the infrared region which is the opposite as showcased in absorption characteristics as the samples with less when compared with the Transmittance. With the 1.0 mole ligand concentration it appears to have less transmittance as in **Figures 2** and **3** respectively which is reflection of the influence of the ligand concentration

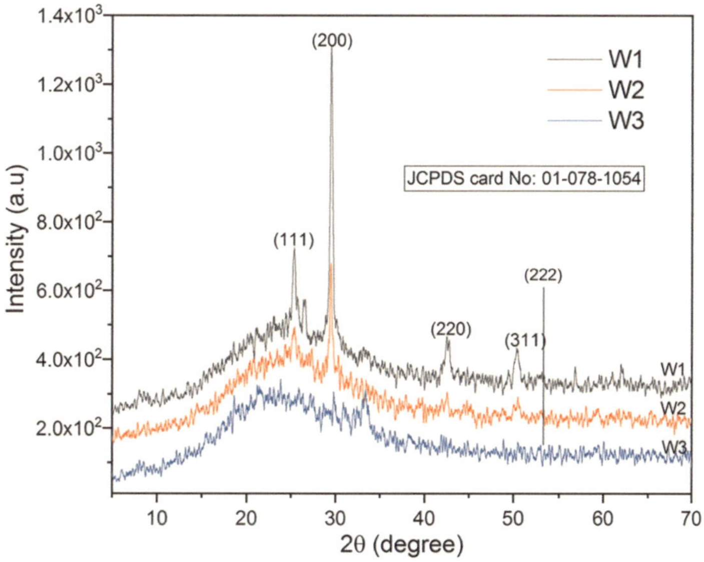

Figure 16.
XRD showing the structural behavior of the PbS thin film for W_1, W_2 and W_3.

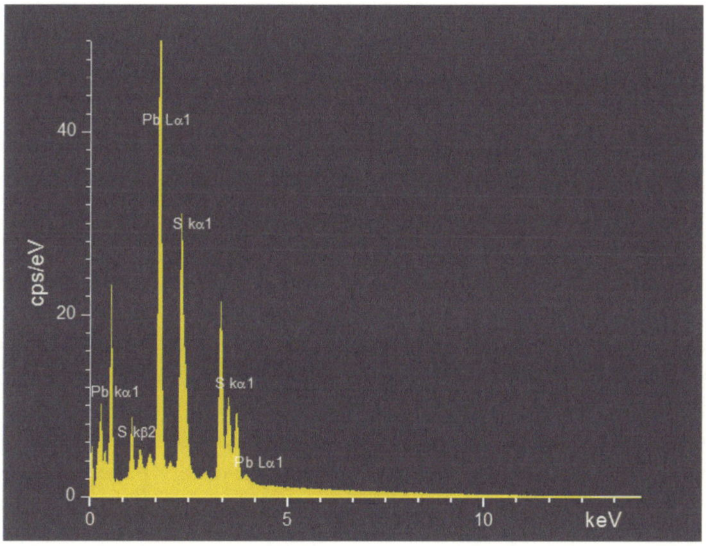

Figure 17.
EDAX spectrum of PbS thin film.

coupled with the effect of the annealing. The energy band gap for the three samples as shown in **Figures 4-6** ranged between 3.35 eV and 4.00 eV indicating that the bad gap widens as the ligand was varied coupled with annealing though PbS material is characterized to be a direct band gap with its good feature in energy conversion that made the thin film good candidate for use in solar cell coupled also with its unique corrosion resistant coating characteristics.

3.2 Structural and morphological feature of PbS thin film

The morphological and structural features are shown in **Figures 13-16** are obtained by SEM and XRD respectively and it is observed that the thin film are covered by spherical grain size that increased with decrease in ligand concentration [19] coupled with the annealing which enhanced spherical orientation of the grain size which as in **Figures 13-15**. The XRD indicates that the thin film has diffraction peak at the plane at (200), (111), (222), (220) and (311) with the preferential prominent peak occurring at (200) with three close lying reflections at (111), (222) and (220) as shown in **Figure 16**. This is as an indication of polycrystalline nature of the film and this varied with the variation of concentration of ligand [20, 21]. The highest intensity occurred at $1.30x10^3 au$ and at 30° and the observation here is that crystalline orientation decreased with the increase in the concentration of the ligand coupled with the annealing effect which has contributed in the enhancement of the grain size as reported by [22]. The elemental composition of the film is shown on **Figure 17** which is the EDAX analysis indicates that only Pb and S are prominent the constituent elements in the in the materials which then culminates the fact that there is no other element present in the thin film and this serves an indication that chemical bath deposition of PbS produced relatively pure thin film of PbS.

At the end of the processes, the samples were annealed at the temperature 80° and the taken to the laboratory for optical, structural and morphological characterization and analysis.

4. Conclusion

The use of varied ligand concentration has been employed to deposit PbS thin film using chemical bath deposition technique under the same time frame and the result presented in this work and from the results, it was observed that there were slight variations in some of the properties as analyzed which have clearly indicated the effect of variation of ligand concentration and the annealing effect on the various properties of the thin film as indicated in figures presented. In a similar manner, it was clearly revealed that the energy band gap of the thin film widened while the transmittance also increased with increase in the concentration of the ligand coupled with obvious indication of variation of grain sizes as indicated in the surface morphology as obtained from the SEM and, XRD and EDAX spectral analysis show that the chemical bath deposited PbS thin film is polycrystalline and relatively pure as it does not contain any impurity element it with little indication of the effect of annealing on the properties of thin film [14]. From the result so far, it could be deduced that generally, good quality of PbS thin film could be grown using CBD with varied ligand because report had already been given that increase in the annealing temperature increases the grain size with enhancement in the strength of the material along with reduction in dislocation density [23, 24]. One of the major attraction in PbS thin film is that the properties of the thin film can be controlled by different deposition conditions such as PH, bath temperature, and ligand concentration just as it is common with so many other chalcogenide based thin films.

Author details

Emmanuel Ifeanyi Ugwu*, Adeyeba Adewale Oluwatobi, Rafiu Adewale Busari, Sunday Ikpughul Lyua, Ibrahim Abdullahi Danasabe, Mohamend Jaafar and Usman Rilwan
Department of Physics, Nigerian University Biu, Nigeria

*Address all correspondence to: ugwuei2@gmail.com; emmanuel.ifeanyi@naub.edu.ng; ugwuei@yahoo.com

References

[1] Onah DU, Ugwu EI, Asiegbu FE. Structural and optical characteristics of cadmium sulphide thin films doped with nickel by solution growth technique. JNAMP. 2013;**24**:377-380

[2] Tejuca LG, Jose LGF, Tascon JMD. Structure and reactivity of perovskite-type oxide. Advance in Catalysis. 1989; **36**:237-238

[3] Priyanka P, Das S, Naik. Various optoelectronic properties for different applications. RSC Advances. 2022;**12**(16): 9599-9620

[4] Soonmin H, Paulraj I, Kumar M, Sonker RK, Nandi P. Recent Developments on the Properties of Chalcogenide Thin Film. London, UK: IntechOpen; 2022. DOI: 10.5772/intechopen.102429. ISBN:978-1-80355-660-4

[5] Greenwood NN, Earnshaw A. Chemistry of the Elements. 2nd ed. Oxford: Butterworth-Heinemann; 1997. ISBN: 0-7506-3365-4

[6] Ugwu EI, Onah DU. Optical characteristics of chemical bath deposited CdS thin film characteristics with UV, visible, and NIR radiation. Pacific Journal of Science and Technology. 2007;**8**(1):155-161

[7] Kadhim AA, Ibrahim A-ME, Marie JM. Structural and optical properties of PbS thin films deposited by pulsed laser deposited (PLD) technique at different annealing temperature. International Journal of Physics. 2017;**5**(1):1-8. DOI: 10.12691/ijp-5-1-1

[8] Wells AF. Structural Inorganic Chemistry. Oxford: Clarendon Press; 1984. ISBN: 0-19-855370-6

[9] Charlie W. Physics Duo Finds Magic in Two Dimensions. Quanta Magazine; 2022 [Accessed: August 22, 2022]

[10] Vaughan DJ, Craig JR. Mineral Chemistry of Metal Sulfides. Cambridge: Cambridge University Press; 1978. ISBN 0521214890

[11] Owoh B, Ugwu EI. Growth and optical characterization of antimony selenide Cu_2S thin film using chemical bath deposition technique. Metallurgy and Material Engineering. 2009;**4**(1):38-41

[12] Nwali NP, Ugwu EI. Optical properties of chemical bath deposited Cu_2S thin film within UV-VIS, and NIR radiation. Metallurgy and Material Engineering. 2009;**4**(2):26-30

[13] Ugwu E. Optical and solid-state properties of manganese sulphide (MnS) thin film, theoretical analysis. International Journal of Multiphysics. 2017;**11**(2):137-150. DOI: 10.21152/1750-9548.11.2.137

[14] Adeyeba AO, Ugwu EI, Adewale B. Influence of ligand on the morphological, electrical, optical and solid state properties chemical bath deposited lead sulphide thin film. Journal of Materials Science and Review. 2022;**10** (4):16-31

[15] Kassim A, Min HS, Wee T, Fei NC. Influence triethanolamine on the chemical bath deposited NiS thin film. American Journal of Applied Science. 2011;**8**(4):356-361

[16] Ugwu EI, John O, Onah DU, John E. Analytical study of the band gap characteristics of copper sulphide thin films: Experiment and computation. AEA International Journal. 2017;**5**(3):42-56

[17] Mane RS, Lokand CD. Chemical bath deposition method for metal chalcogenides thin film materials. Chemistry and Physics. 2000;**65**:1-31

[18] Sze SM, Kowok KN. Physics of semiconductor. Wiley Interdisciplinary. 2006;**31**:918-920

[19] Fan DB, Wang H, Zhang YC, Cheng J, Wang B, Yan H. Preparation of crystalline MnS thin film by chemical bath deposition. Materials Chemical Physics. 2003;**80**(1):44-47

[20] Deshpande MP, Bhatt GN, Sakriya P, Chariya P, Chaki SH. Characterization of CdSe thin film deposited by chemical bath solution containing thriethanolamine. Materials Science Semiconductor Proceedings. 2013;**16**(3): 915-220

[21] Chaki SH, Chauhan SM, Tailor JP, et al. Synthesis of manganess sulfide (MnS) thin film by chemical bath deposition and their characterization. Journal of Material Research and Technology. 2013; **6**(2):123-128

[22] Mousa AM, Ali SB. Effects of deposition parameters on chemically deposited PbS thin film. Iraqi Journal of Applied Physics. 2008;**4**(4):7-11

[23] Oriaku CI, Osuwa JC. Analysis of thin chalcogenide PbS films prepared from chemical bath. PJST. 2008;**9**:460-467

[24] Prem Kumar T, Saravanakumar S, Sankaranaryanan K. Effects of annealing on the surface and band gap alignment of CdZnS thin films. Applied Surface Science. 2011;**257**:1923-1927

Chapter 2

Electronic and Optical Properties of Multilayer $PtSe_2$

Nour El Houda Safi

Abstract

Two-dimensional materials arouse ever greater interest in the scientific community due to their electronic and optical properties. Among these 2D materials, the 2D family of transition metal dichalcogenides (TMDs) offers great potential for applications in optoelectronics and nanotechnology. Of these TMD nanomaterials, platinum diselenide $PtSe_2$ has been extensively studied since the successful synthesis of a $PtSe_2$ monolayer in 2015. In this chapter, the multilayer $PtSe_2$ is investigated with first-principle calculations. In order to calculate the optical properties of the system, we first determine its dielectric function. From this, we can extract other optical functions, such as refractive index, extinction coefficient, absorption coefficient, and reflectivity. A good description of these properties can be enhanced by a detailed study of the material's band structure.

Keywords: platinum diselenide, transition metal dichalcogenides (TMDs), DFT, quantum confinement, band structures calculations, optical properties

1. Introduction

In 2004, Konstantin Novoselov and André Geim succeeded in exfoliating graphene from graphite. For this, they were awarded the Nobel Prize in Physics in 2010. Since this discovery, the scientific community has become very interested in 2D materials, due to graphene's physical properties. Indeed, this one has shown high charge carrier mobility of $2.10^5 cm^2/Vs$, excellent flexibility, transparency of 95% [1], high electrical conductivity, and thermal conductivity equal to $10^4\ \Omega^{-1}\ cm^{-1}$ and $3000\ W/mK$ [2], respectively. However, graphene is a zero-gap semimetal. This limits its use in optoelectronics. Other similar 2D materials, such as transition metal dichalcogenide semiconductors (TMDs), are paving the way for optoelectronic applications [3].

This present work deals with one of the TMDs $PtSe_2$. It has a wide range of applications such as photodetectors [4, 5], gas sensing [6], and electronics. Platinum diselenide $PtSe_2$ belongs to the transition metal dichalcogenide (TMD) family. Compared with other materials in the TMDs family, $PtSe_2$ is characterized by high electron mobility at room temperature. The platinum diselenide material $PtSe_2$ is centrosymmetric, stable in the 1T phase, with symmetry group D_{3d} (space group $P\bar{3}m1$), and crystallizes under the hexagonal structure [7]. The primitive cell of $PtSe_2$ contains three atoms: one platinum (Pt) and two selenium (Se). In the same plane, the atoms are connected by covalent bonds with no charge transfer between them, whereas, out of the plane, the Se-Se atoms interact only through weak van der Waals (vdw) interaction.

The real lattice vectors are given by [8]:

$$\begin{cases} \vec{a} = 0.5 \quad a\left(3\vec{x} - \vec{y}\right) \\ \vec{b} = 0.5 \quad a\left(3\vec{x} + \vec{y}\right) \end{cases} \tag{1}$$

Figure 1b shows a $PtSe_2$ monolayer where the lattice parameter a is of order 3.7° A, the Pt-Se bond length is 2.52° A and the distance noted 'h' separating the $Se_{sup} - Se_{inf}$ atoms is equal to 2.53° A. These parameters are measured using a scanning tunneling microscope (STM) [10].

2. Electronic properties

The determination of a large band gap for multilayer $PtSe_2$ remains a major theoretical and experimental challenge. At the experimental level, a study [11] showed by scanning tunneling spectroscopy (STS) and scanning tunneling microscopy (STM), that the value of the gap energy decreases from a value of 1.79 pm 0.04 eV for monolayer to 0.62 pm 0.02 eV for bilayer, and then, it cancels for trilayer. In contrast, another literature [12] showed that the trilayer $PtSe_2$ is a semiconductor with a gap of 0.2 eV.

At the theoretical level, studies such as GW overestimate the gap energy, which predict that the band gap of monolayer, bilayer, and trilayer $PtSe_2$ are equal,

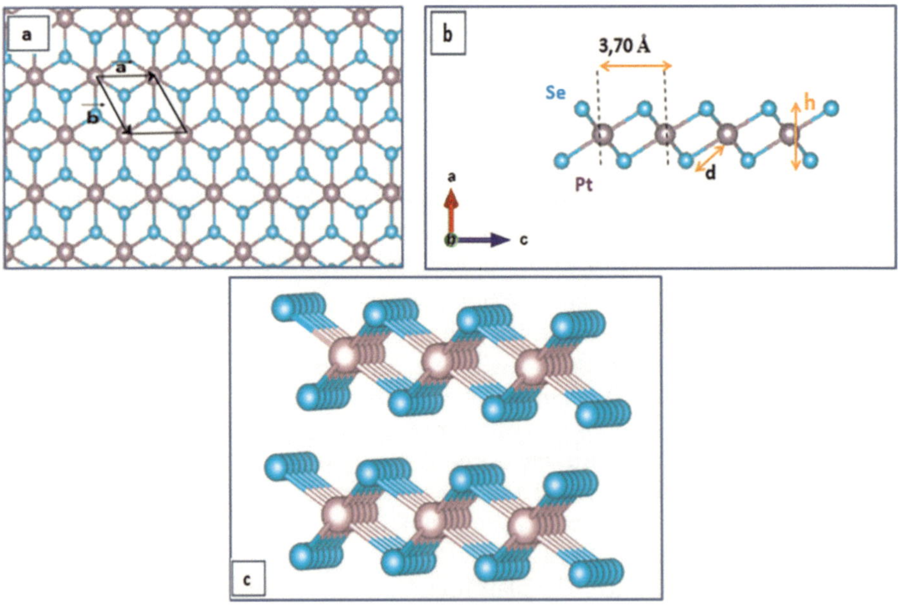

Figure 1.
Structure of $PtSe_2$ (a) top view where the elemental lattice is presented by a rhombus, (c) side view, atoms in blue; Se and atoms in purple; Pt, (b) monolayer $PtSe_2$, visualized by Vesta [9].

respectively, to 2.4, 1.5, and 0.7 eV. However, other theories based on the density functional approach give different estimates due to the accuracy of the adapted exchange-correlation energy functional. For example, the standard density functional theory (DFT) approximations (GGA-LDA) underestimate the gap energy, and they predict that the gap energy of the monolayer, bilayer and trilayer $PtSe_2$ are equal, respectively, to: 1.2, 0.2, and 0 eV [12]. On the contrary, the hybrid functional HSE06 shows excellent accuracy in determining the prohibited bandwidths for the $PtSe_2$ monolayer, bilayer, and trilayer: 1.7 (eV), 0.52 (eV) and 0.13 (eV) [13]. However HSE06 requires a large memory capacity and a very long computation time.

2.1 Calculation method

The electronic and optical properties are determined by the Quantum Espresso (QE) code [14]. The Perdew-Burke-Ernzerhof (PBE-GGA) [15] and hybrid functionals optB86b-vdw [16] are adopted for the exchange-correlation potential. The exchange-correlation energy (XC) between electrons is processed by the XC functional. The interaction between electrons and their nucleus is described by the projection-augmented plane wave (PAW) method, in which the valence electrons used are 10 platinum (Pt) $(5d^9 6s^1)$ and 6 selenium (Se) $(4s^2 4p^4)$. In order to obtain reliable and similar results to the different experimental and theoretical literatures, we proceed to choose appropriate values for specific important parameters: the cut energy E_{cut} and the number of k-points in the Monkhorst-Pack grid in the Brillouin zone (ZB) [2] according to a convergence criterion. Based on these tests, we applied 60 Ry for the cutoff energy and $15 \times 15 \times 1$ for k-points for the relaxation and self-consistent calculation for multilayer $PtSe_2$. On the contrary, the bulk $PtSe_2$ is treated with E_{cut} equal to 90 Ry and k-points $15 \times 15 \times 15$. In order to calculate the density of states, the cutoff energy is set at 90 Ry for all systems, and the k-points in the ZB chosen for the monolayer and bilayer of $PtSe_2$ are $40 \times 40 \times 1$, while those for the trilayer and quadrilayer are $33 \times 33 \times 1$. For the bulk $PtSe_2$, the k-points are kept at $33 \times 33 \times 33$ (**Figure 2**).

By using the *«Variable Cell-relax»* option of Quantum Espresso that allows relaxing both ions and the crystal lattice, the structure of the system is relaxed up to the total energy is less than 10^{-5}Ry and the forces acting on atoms are less than 10^{-4}Ry/bohr^3. To account for van der Waals-type interactions, we used the Gimmes DT-D3(BJ)

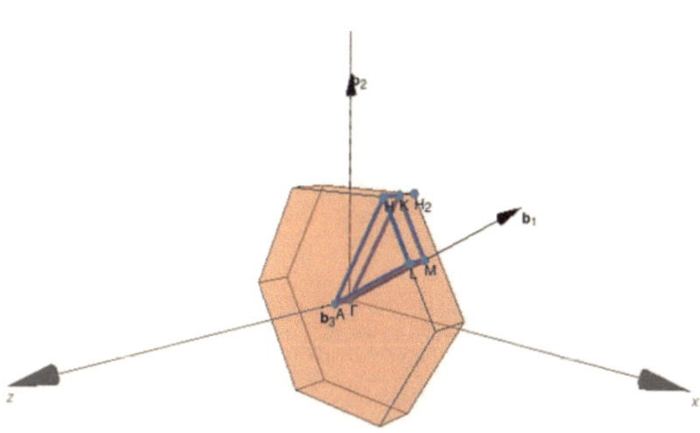

Figure 2.
Brillouin zone, the blue triangle is the irreducible Brillouin zone of the monolayer $PtSe_2$.

correction. We add a vacuum space on the z axis to avoid interaction with the neighboring mesh in the z direction.

2.1.1 Band structure of $PtSe_2$ monolayer, multilayer, and bulk

The band structure calculation for multilayer $PtSe_2$ are performed along the $K - \Gamma - M - K$ path in the irreducible Brillouin zone. It is well known that PBE functionals underestimate bandgaps. We thus add the van der Waals correction, which is close to the experimental result. However, PBE-D3 predicts the semiconductor–metal transition of $PtSe_2$ films from three layers, which disagrees with experimental data [12]. Therefore, we adopt the optB86b functional [16] for further calculations for bilayer and trilayer $PtSe_2$.

2.1.1.1 Monolayer $PtSe_2$

The $PtSe_2$ monolayer is a semiconductor with an indirect gap of 1.448 eV without spin-orbit coupling (SOC) and 1.259 eV with SOC, which is in good agreement with previous theoretical studies of the same system [17, 18]. The conduction band minimum is located between the points "Γ" and "M," and the valence band is at the point "Γ." Taking into account the spin-orbit coupling effect, we notice that the degeneration at least of the valence band point "Γ" has disappeared, and the gap energy decreases (**Figure 3**).

2.1.1.2 Bilayer $PtSe_2$

The $PtSe_2$ bilayer is a semiconductor with an indirect gap of 0.14 eV by PBE-D3 functional and 0.30 eV by optB86b-vdw functional. OptB86b-vdw functional gives a good agreement with previous theoretical studies [17]. The conduction band minimum is between Γ and M points, analogous to the band structure of the monolayer. For comparison to monolayer, the valence band maximum is translated between the points K-Γor Γ-M (**Figure 4**).

2.1.1.3 Trilayer $PtSe_2$

The optB86b functional correctly predicts the semiconductor bandgap of the trilayer (0.04 eV) as observed in experiments [17]. The conduction band minimum is between the Γ point and M, and the valence band maximum is in Γ point (**Figure 5**).

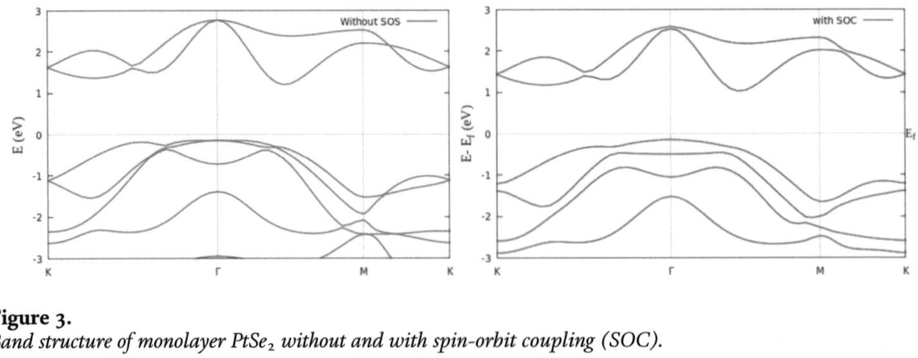

Figure 3.
Band structure of monolayer $PtSe_2$ without and with spin-orbit coupling (SOC).

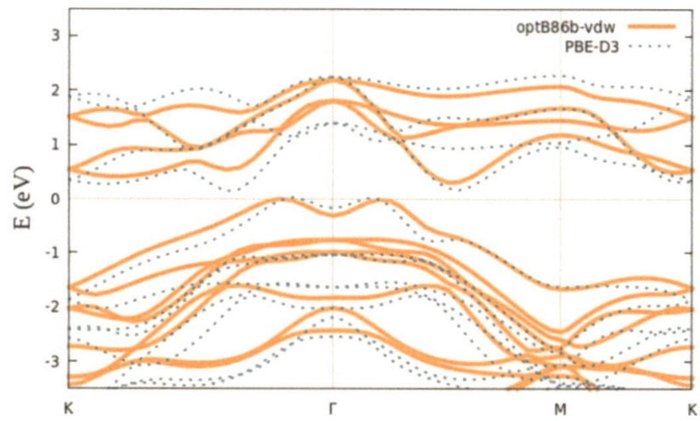

Figure 4.
Band structure of bilayer PtSe$_2$.

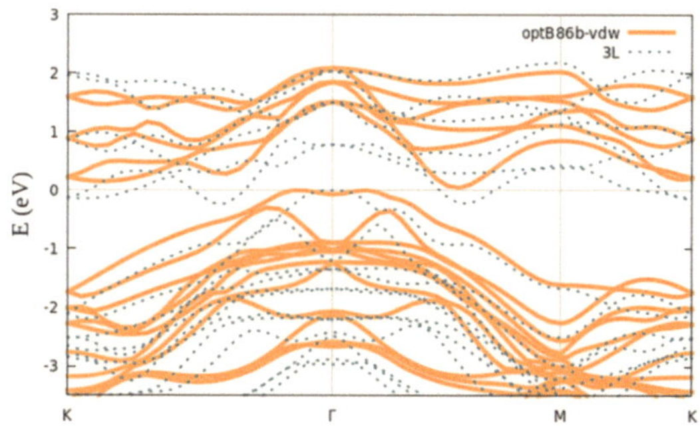

Figure 5.
Band structure of trilayer PtSe$_2$.

2.1.1.4 Quadrilayer $PtSe_2$

See **Figure 6**.

The conduction and valence bands began to overlap ($E_c - E_v = -0.56\,eV$ with E_c is the minimum of the conduction band energy and E_v is the maximum of the valence band energy) which show that the quadrilayer of $PtSe_2$ is a semi-metal. The conduction band minimum is between the Γ point and M, and the valence band maximum is in Γ points, similar to monolayer and trilayer $PtSe_2$ films.

2.1.1.5 Bulk $PtSe_2$

The bulk $PtSe_2$ is a semi-metal Dirac type II [19], represented under the hexagonal structure (**Figure 7**):

The band structure of bulk $PtSe_2$ exhibited a semimetallic band structure along the Γ-A path. The conduction band minimum is at point K or between points H and A, and the valence band maximum is in points Γ. The conduction and valence bands began to overlap ($E_c - E_v = -0.58 \quad eV$ with E_c is the minimum of the conduction band energy and E_v is the maximum of the valence band energy) (**Figure 8**).

2.1.2 Densities of states of $PtSe_2$ monolayer, multilayer and bulk

Figures 9–13 represent the total and partial densities of states of multilayer $PtSe_2$ films.

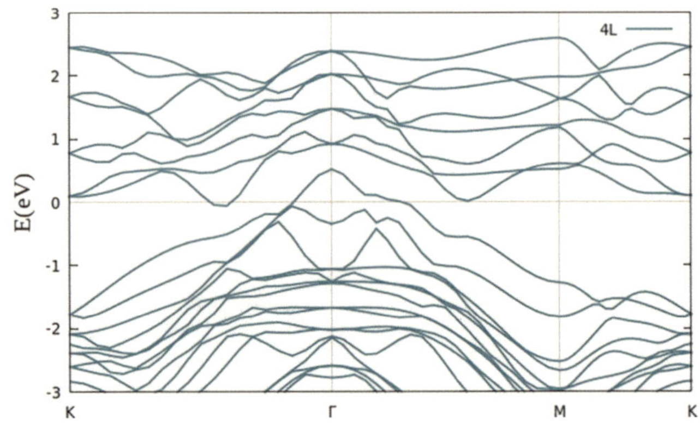

Figure 6.
Band structure of quadrilayer $PtSe_2$.

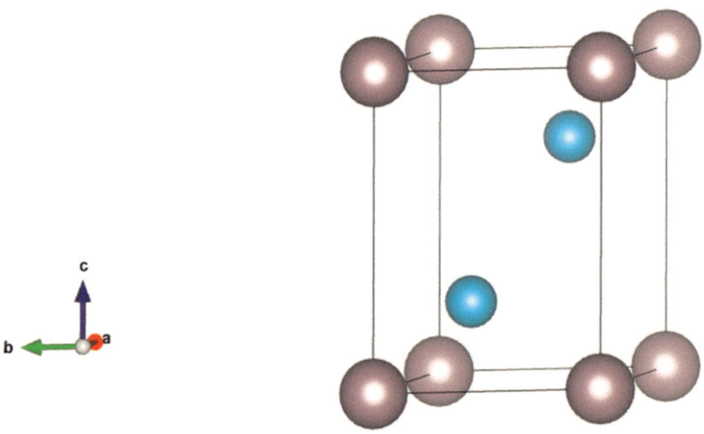

Figure 7.
Bulk $PtSe_2$ structure in lateral view, visualized by VESTA [9]. Atoms in blue are: Se and atoms in purple are: Pt.

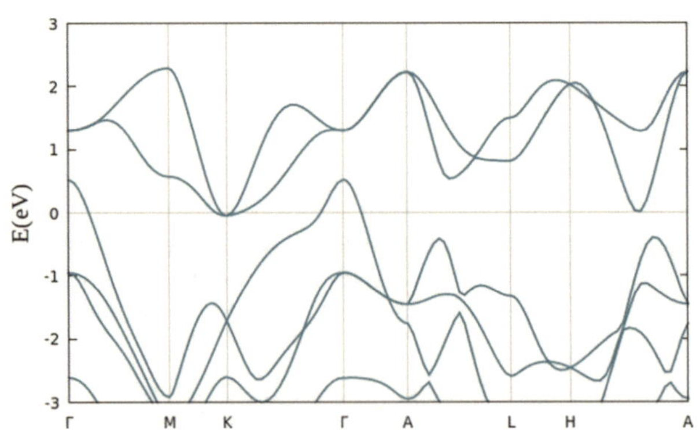

Figure 8.
Band structure of bulk $PtSe_2$.

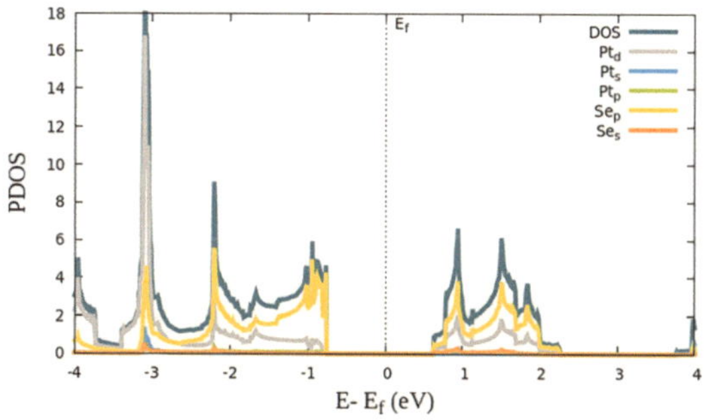

Figure 9.
Calculated partial density of states (PDOS) for monolayer $PtSe_2$ *films.*

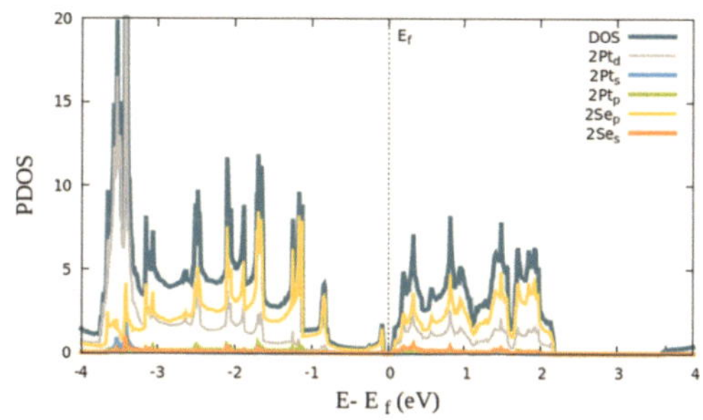

Figure 10.
Calculated partial density of states (PDOS) for bilayer $PtSe_2$ *films.*

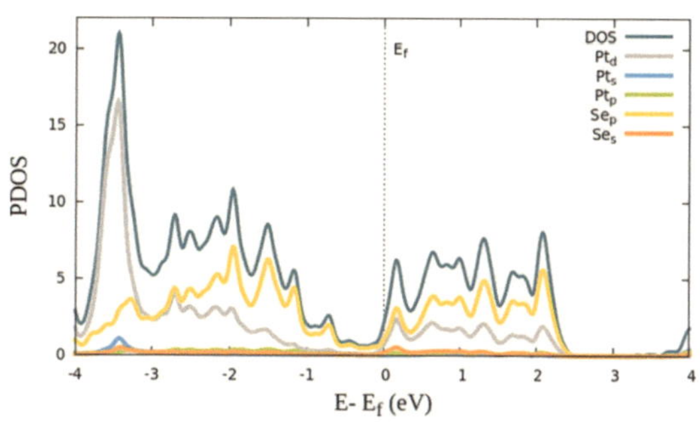

Figure 11.
Calculated partial density of states (PDOS) for trilayer $PtSe_2$ *films.*

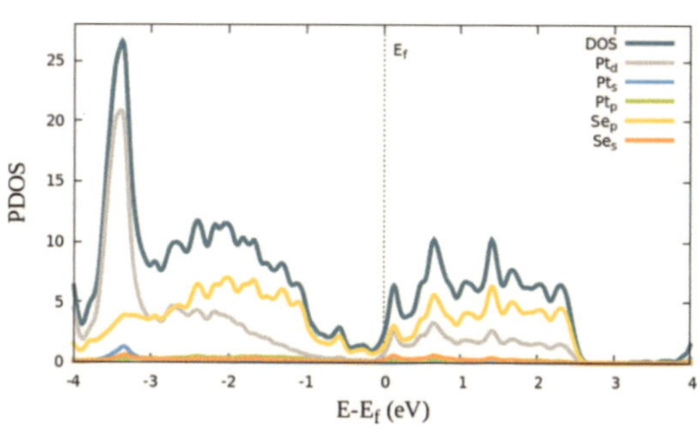

Figure 12.
Calculated partial density of states (PDOS) for quadrilayer $PtSe_2$ *films.*

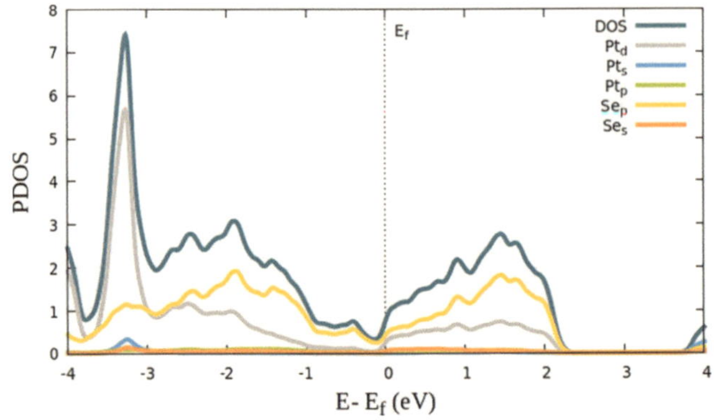

Figure 13.
Calculated partial density of states (PDOS) for bulk $PtSe_2$ films.

According to the PDOS structure, the platinum (Pt) and selenium (Se) p-orbitals are strongly hybridized and are responsible for valence band and conduction band formation. The density of electronic states clearly shows Van Hove singularities (peaks). These critical points provide information on the different direct transitions between the valence and conduction bands [20]. From monolayer to bulk $PtSe_2$, the intensity of these peaks decreases and the density structure becomes more smoother. Then, these Van Hove singularities are the result of the two-dimensional character of $PtSe_2$.

3. Optical properties

3.1 Dielectric function

When an electromagnetic wave propagates in a dispersive material medium at frequencies of the same order of magnitude as the electronic vibration frequencies of this material, the real dielectric constant ε becomes a function of the energy of incident photons. Thus, if this medium is absorbent, the dielectric function is written in the complex form $\varepsilon(\omega)$ [21].

3.2 Calculation method

The dielectric function was calculated by the codeturboEELS [22] that is a component of Quantum Espresso [14, 23, 24]. This code adapts the theory of perturbation of the time-dependent density functional (TDDFPT) based on the Liouville-Lanczos (LL) approach. This method has been established for electron energy loss spectroscopy (EELS) that is directly related to the dielectric response of the material. Moreover, this approach makes it possible to calculate the dielectric function for any value of the transfer moment q. For small q, long-range effects are important, while for large q short-range effects are dominant. This approach has several advantages over the conventional TDDFT approach, such as no counted empty state, which allows to extend the EELS spectra calculations from the low-loss region to the 50–100 eV region [25].

The calculation was determined by the use of Norm-conserving pseudo-potential, the GGA approximation for exchange-correlation energy (XC) and for weak photon wave vector values q. The incident field is modeled only under the direction x. We consider only *interband radiative transitions*.

3.2.1 Imaginary part of the dielectric function

The imaginary part of the dielectric function of multilayer $PtSe_2$ films is shown in **Figure 14**.

The imaginary part $Im(\varepsilon)$ of $PtSe_2$ is strictly positive. It is important in the incident photon energy region [1–6]. Also, $Im(\varepsilon)$ rises as a function of incident photon energy toward a dominant peak, and then, it decreases to large energy values. Noting that by increasing the thickness of the material $PtSe_2$, this peak becomes more and more intense at lower energy value (red shift). The following table shows the variation of this peak of interband transitions from the monolayer to the bulk $PtSe_2$ (**Table 1**).

The imaginary part of the dielectric function $Im(\varepsilon)$ provides information on the absorption of the medium. According to **Figure 14**, the absorption tends to increase when the dimension of $PtSe_2$ is high. As a result, the $PtSe_2$ massif is characterized by high optical absorption compared to $PtSe_2$ 2D materials. In addition, the $Im(\varepsilon)$ is proportional to the sum of all transitions between the valence and conduction bands. As mentioned, the platinum orbital d (Pt) and the selenium orbital p (Se) are responsible for the formation of valence and conduction bands. Therefore, the spectrum of $Im(\varepsilon)$ reflects three critical peaks corresponding to the transitions between the orbitals d(Pt) and p(Se).

3.2.2 Real part of the dielectric function

See **Figure 15**.

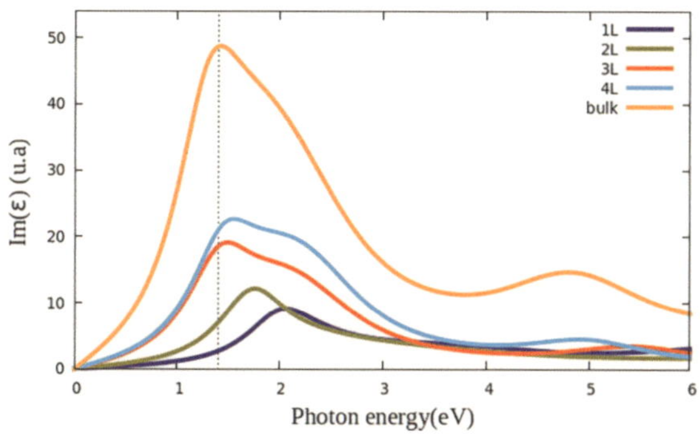

Figure 14.
Variation of the imaginary part of the dielectric function as a function of energy for multilayer $PtSe_2$.

$1T-PtSe_2$	Monolayer	Bilayer	Trilayer	Quadrilayer	Bulk
Photon energy (eV)	2.4	1.761	1.5	1.465	1.422
$Im(\varepsilon)$ (u.a)	7.9	12.286	19.189	28	48.749

Table 1.
Maximum peak variation $Im(\varepsilon)$ *from the monolayer to the bulk* $PtSe_2$.

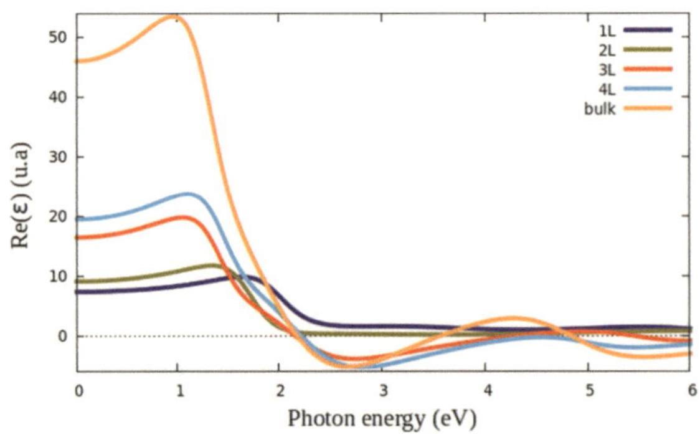

Figure 15.
Variation of the real part of the dielectric function for multilayer $PtSe_2$.

$1T - PtSe_2$	Monolayer	Bilayer	Trilayer	Quadrilayer	bulk
$\varepsilon[\omega = 0]$ (u.a)	6.07	9.07	16.537	24.55	46
Other literatures	6.71 [26]				40 [27]

Table 2.
Variation of static ε going from the monolayer to the bulk $PtSe_2$.

The spectrum, below, shows the decrease of the real part of the dielectric function $Re\,(\varepsilon)$ of the material $PtSe_2$ by increasing the photon energy. From zero eV, $Re\,(\varepsilon)$ starts with positive values going toward a maximum, then it decreases toward zero for semiconductors: 1L and 2L or to a minimum of −5 (u.a) for semi-metals: 3L, 4L, and bulk. Then it is followed by oscillations near zero in the ultraviolet (UV) region. In addition, $Re\,(\varepsilon)$ tends to have a maximum at lower energy values from the monolayer to the bulk $PtSe_2$. From these figures, we can extract the values of the static dielectric constant $\varepsilon[\omega = 0]$ which represents the dielectric response to the static electric field of $PtSe_2$. These values are grouped in **Table 2**.

This variation of the constant $\varepsilon(0)$ can be explained by the Penn model [28]. According to Eq. (2), the dielectric constant decreases by increasing the gap energy:

$$\varepsilon = 1 + \frac{h\omega}{\mathcal{E}_g} \quad (2)$$

In addition, according to the spectrum [15], the $Re\,(\varepsilon)$ of the trilayer, quadrilayer, and bulk $PtSe_2$ have negative values. This means that the electromagnetic wave does not propagate in the semi-metals 3L, 4L, and the bulk $PtSe_2$, and that the reflection and absorption processes occurs.

The shift of the maximum peaks of the $Re\,(\varepsilon)$ and $Im(\varepsilon)$ to very low energy values (red shift) can be explained by the quantum confinement effect that is the result of the material shrink. Moreover, under an external excitation, the photon energy emitted by nano-objects depends on their size, with increasing size, the emission tends to lower energies [29]. This is the quantum confinement effect.

3.2.3 Comparison with theoretical studies

The literatures [26, 27] calculated the linear optical response ε for the monolayer and the bulk $PtSe_2$ using WIEN2K and VASP software, respectively. The dielectric functions determined for the monolayer and the bulk $PtSe_2$ in the [27] study are similar to our calculations. The $Im(\varepsilon)_{bulk}$ shows a very strong maximum at an energy value of 1.25 that is close to the energy value calculated by the QE code at 1.4 eV for a maximum of $Im(\varepsilon)$. The $Re\,(\varepsilon)_{bulk}$ has a very intense maximum at an energy value of 1 eV [27] that is very close to the energy value found by the QE code of 0.91 eV for maximum intensity.

3.3 Refractive index and extinction coefficient

The dielectric function is a fundamental physical parameter to study the optical properties of the material medium. But often, opticians adapt other functions are easier for a better description of properties, such as the refractive index used in the field of photovoltaics and the absorption coefficient for transparent materials.

The complex refractive index N is directly related to the dielectric function ε by the relation $N^2 = \varepsilon$, where N is defined by:

$$N = n - i \quad \mathcal{K} \tag{3}$$

With $n(\omega)$ is the refractive index and $\mathcal{K}(\omega)$ the extinction coefficient. So, the real part and imaginary ε are written:

$$Re\,(\varepsilon[\omega]) = n(\omega)^2 - \mathcal{K}(\omega)^2 \quad \text{et} \quad Im(\varepsilon[\omega]) = 2n\mathcal{K} \tag{4}$$

We calculate the refractive index $n(\omega)$ and the extinction coefficient $\mathcal{K}(\omega)$ using the following expressions [21] :

$$n(\omega) = \left(\frac{\sqrt{Re\,(\varepsilon[\omega])^2 + Im(\varepsilon[\omega])^2}}{2} + \frac{Re\,(\varepsilon[\omega])}{2} \right)^{\frac{1}{2}} \tag{5}$$

$$\mathcal{K}(\omega) = \left(\frac{\sqrt{Re\,(\varepsilon[\omega])^2 + Im(\varepsilon[\omega])^2}}{2} - \frac{Re\,(\varepsilon[\omega])}{2} \right)^{\frac{1}{2}} \tag{6}$$

See **Figures 16** and **17**.

The refractive index curve shows a light shift analogous to the dielectric function as a function of thickness. In fact, the maximum peaks of $n(\omega)$ tend to distribute toward low energies by increasing the size of multilayer $PtSe_2$. As previously indicated, this displacement is a consequence of quantum confinement phenomenon due to the thickness effect. From the spectra 16, the static refractive index can be extracted (**Table 3**).

The static refractive index measures the superiority of the speed of light c in a vacuum (or air) compared to over that in the material: $n = \frac{c}{v}$. The calculated refractive index values are in good agreement with theoretical results from other literature. Moreover, they respect the relation [31] $n(0) = \sqrt{\varepsilon(0)}$.

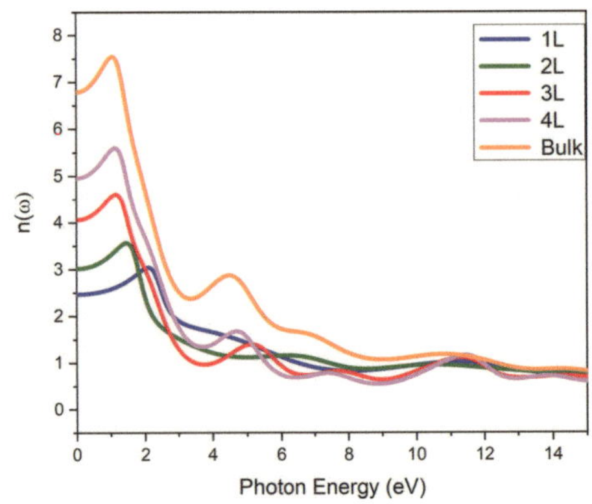

Figure 16.
Variation of refractive index as a function of energy for $PtSe_2$ multilayer.

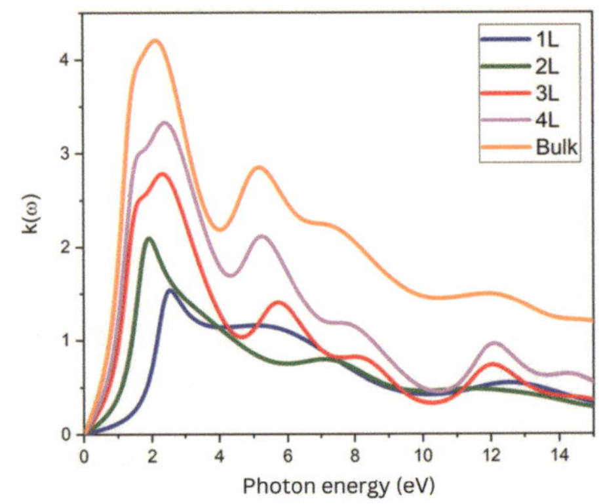

Figure 17.
Variation of extinction coefficient as a function of energy for $PtSe_2$ multilayer.

$1T - PtSe_2$	Monolayer	Bilayer	Trilayer	Quadrilayer	Bulk
$n(0)$	2.449	3.016	3.99	4.956	6.78
Other literatures	2.95 [26], 2.08 [30]				

Table 3.
Variation of static refractive index as a function of energy for $PtSe_2$ multilayer.

The extinction coefficient measures the fraction of light lost due to scattering and absorption at a particular wavelength: $\mathcal{K} = \frac{\alpha\lambda}{4\pi}$.

The spectrum of $\mathcal{K}(\omega)$ [17] reveals an increase in the infrared for the bilayer, trilayer, quadricayer, and bulk $PtSe_2$. In contrast, the extinction coefficient of the monolayer $PtSe_2$ begins to increase in the visible region. In the ultraviolet region, the extinction coefficient of multilayer $PtSe_2$ decreases until it stabilizes near 1.

3.4 Reflectivity and absorption coefficient

The α absorption coefficient measures the energy loss of electromagnetic radiation as it passes through the absorbing medium. This coefficient is calculated from the expression:

$$\alpha(\omega) = 4\pi \quad \frac{\mathcal{K}(\mathcal{E})}{\lambda} = \frac{4\pi}{hc} \quad \mathcal{K}. \quad \mathcal{E} \tag{7}$$

Reflectivity is determined by the following formula [21]:

$$R(\omega) = \left|\frac{N-1}{N+1}\right|^2 = \frac{(n-1)^2 + \mathcal{K}^2}{(n+1)^2 + \mathcal{K}^2} \tag{8}$$

According to these curves, platinum diselenide $PtSe_2$ is a good absorber in the visible and ultraviolet (UV) region since the $PtSe_2$ bulk shows a strong absorption with a maximum value equal to 1.8 $10^6 cm^{-1}$ in the ultraviolet (UV) range.

As seen in **Figure 18**, multilayer $PtSe_2$ shows transparency in the infrared range. However, the $PtSe_2$ bulk has a very high reflectivity, covering the entire spectrum. In contrast, the reflectivity of $PtSe_2$ in multilayers is much lower than that of the bulk, tending to zero at 10 eV. In addition, the bilayer and monolayer are less reflective than the other systems. The reflectivity of the bilayer and monolayer starts to decrease at 1.8 eV and 2.3 eV, respectively. As a result, multilayer $PtSe_2$ is more absorbent than bulk $PtSe_2$. This gives great potential for optoelectronic applications in the visible and ultraviolet range (**Figure 19**).

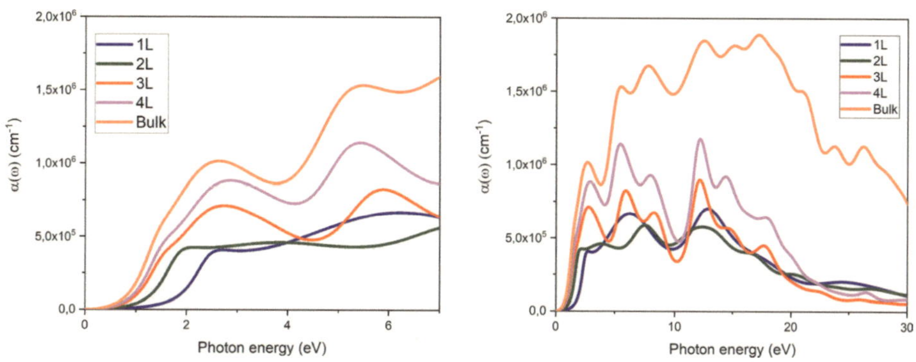

Figure 18.
Variation of the absorption coefficient as a function of energy for multilayer PtSe$_2$ in the energy range [0–6] eV and [0–30] eV.

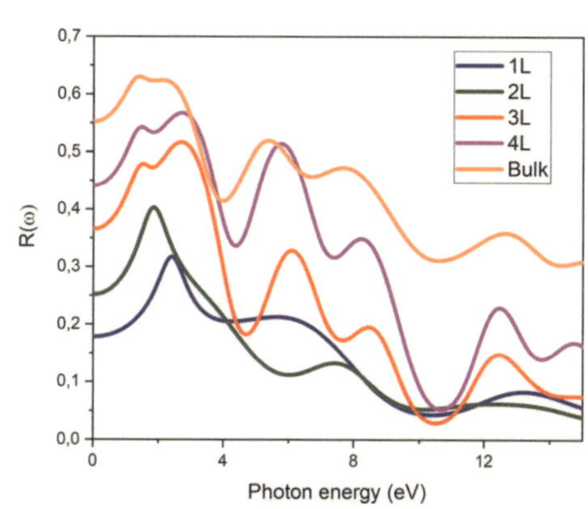

Figure 19.
Variation of reflectivity as a function of energy for multilayer PtSe$_2$.

4. Conclusion

In this chapter, we have studied the electronic and optical properties of multilayer $PtSe_2$. This study is based on density functional theory (DFT) and its TDDFT extension. Our results are in good agreement with both theoretical and experimental literature.

This work presents a theoretical study of the effect of thickness on the properties of $PtSe_2$ in multilayers, such as: electronic band structure, density of states, dielectric function, extinction coefficient, refractive index, absorption coefficient, and reflectivity. Firstly, we have shown the transition of multilayer platinum diselenide $PtSe_2$ from a semimetal (in bulk structure) to a semiconductor (in monolayer structure). We have adapted the hyride exchange and correlation functional (XC): optB86b-vdw to describe correctly the transition from the semiconducting to the semimetallic state from the fourth layer. Secondly, we studied the total and partial densities of states to identify the contributions of orbitals in the valence and conduction bands. The variation of the density of states as a function of the number of layers of $PtSe_2$ shows Van Hove singularities due to the two-dimensional character of $PtSe_2$. Next, we determined the influence of thickness on the optical properties of $PtSe_2$ multilayers.

This thickness dependence of electronic and optical properties is the sign of quantum confinement in multilayer $PtSe_2$ nanoparticles. In addition, we investigated the absorption coefficient, reflectivity, refractive index, and extinction coefficient of multilayer $PtSe_2$. We conclude that multilayer $PtSe_2$ is more absorbent than bulk, making it a promising candidate for optoelectronic applications.

A promising prospect of the multilayer platinum $PtSe_2$ is the determination of its nonlinear optical properties to investigate the possibility of a THz emission.

Author details

Nour El Houda Safi
Department of Physics, Tunis Faculty of Science, Tunis El Manar University, Tunisia

*Address all correspondence to: nourelhoudasafi07@gmail.com

References

[1] Wang X, Zhi L, Müllen K. Transparent, conductive graphene electrodes for dye-sensitized solar cells. Nano Letters. 2008;**8**(1):323-327

[2] Allen MJ, Tung VC, Kaner RB. Honeycomb carbon: A review of graphene. Chemical Reviews. 2010; **110**(1):132-145

[3] Jenaina Ribeiro-Soares RM, Almeida EB, Barros PT, Araujo MS, Dresselhaus LG, Cançado AJ. Group theory analysis of phonons in two-dimensional transition metal dichalcogenides. Physical Review B. 2014;**90**(11):115438

[4] Di W, Wang Y, Zeng L, Jia C, Enping W, Tingting X, et al. Design of 2d layered ptse2 heterojunction for the high-performance, room-temperature, broadband, infrared photodetector. ACS Photonics. 2018;**5**(9):3820-3827

[5] Wang G, Wang Z, McEvoy N, Fan P, Blau WJ. Layered ptse2 for sensing, photonic, and (opto-) electronic applications. Advanced Materials. 2021; **33**(1):2004070

[6] Yim C, Lee K, McEvoy N, O'Brien M, Riazimehr S, Berner NC, et al. High-performance hybrid electronic devices from layered ptse2 films grown at low temperature. ACS Nano. 2016;**10**(10): 9550-9558

[7] Chen J, Wang Y, Wenshuo X, Wen Y, Ryu GH, Grossman JC, et al. Atomic structure of dislocations and grain boundaries in two-dimensional ptse2. ACS Nano. 2021;**15**(10):16748-16759

[8] Ali Kandemir B, Akbali ZK, Badalov SV, Ozcan M, İyikanat F, Sahin H. Structural, electronic and phononic properties of ptse2: From monolayer to bulk. Semiconductor Science and Technology. 2018;**33**(8): 085002

[9] Momma K, Izumi F. Vesta 3 for three-dimensional visualization of crystal, volumetric and morphology data. Journal of Applied Crystallography. 2011;**44**(6):1272-1276

[10] Wang Y, Li L, Yao W, Shiru Song JT, Sun JP, Ren X, et al. Monolayer ptse2, a new semiconducting transition-metal-dichalcogenide, epitaxially grown by direct selenization of pt. Nano Letters. 2015;**15**(6):4013-4018

[11] Li J, Kolekar S, Ghorbani-Asl M, Lehnert T, Biskupek J, Kaiser U, et al. Layer-dependent band gaps of platinum dichalcogenides. ACS Nano. 2021;**15**(8): 13249-13259

[12] Xiong K, Hilse M, Li L, Göritz A, Lisker M, Wietstruck M, et al. Large-scale fabrication of submicrometer-gate-length mosfets with a trilayer ptse 2 channel grown by molecular beam epitaxy. IEEE Transactions on Electron Devices. 2020;**67**(3):796-801

[13] Dake H, Zhao T, Ping X, Zheng H, Xing L, Liu X, et al. Unveiling the layer-dependent catalytic activity of ptse2 atomic crystals for the hydrogen evolution reaction. Angewandte Chemie. 2019;**131**(21):7051-7055

[14] Giannozzi P, Baroni S, Bonini N, Calandra M, Car R, Cavazzoni C, et al. Quantum espresso: A modular and open-source software project for quantum simulations of materials. Journal of Physics: Condensed Matter. 2009;**21**(39): 395502

[15] Perdew JP, Burke K, Ernzerhof M. Generalized gradient approximation

made simple. Physical Review Letters. 1996;**77**(18):3865

[16] Klimeš J, Bowler DR, Michaelides A. Chemical accuracy for the van der waals density functional. Journal of Physics: Condensed Matter. 2009;**22**(2):022201

[17] Fang L, Liang W, Feng Q, Luo S-N. Structural engineering of bilayer ptse2 thin films: A first-principles study. Journal of Physics: Condensed Matter. 2019;**31**(45):455001

[18] Manchanda P, Kumar P, Dev P. Defect-induced 4 p-magnetism in layered platinum diselenide. Physical Review B. 2021;**103**(14):144403

[19] Zhang K, Yan M, Zhang H, Huang H, Arita M, Sun Z, et al. Experimental evidence for type-ii dirac semimetal in ptse 2. Physical Review B. 2017;**96**(12): 125102

[20] Mansour M. Étude des propriétés optiques de nanoparticules de semiconducteurs: Application de l'ellipsométrie aux silicium et germanium nanostructurés [PhD thesis]. Université Paul Verlaine-Metz; 2006

[21] Keita AS. Etude par ellipsométrie spectroscopique des effets de taille sur les propriétés optiques de couches composites à matrice diélectrique et du silicium nanostructuré [PhD thesis]. Université de Lorraine; 2012

[22] Timrov I, Vast N, Gebauer R, Baroni S. Turboeels—a code for the simulation of the electron energy loss and inelastic x-ray scattering spectra using the liouville–lanczos approach to time-dependent density-functional perturbation theory. Computer Physics Communications. 2015;**196**:460-469

[23] Giannozzi P, Andreussi O, Brumme T, Oana Bunau M, Nardelli B, Calandra M, et al. Advanced capabilities for materials modelling with quantum espresso. Journal of Physics: Condensed Matter. 2017;**29**(46):465901

[24] Malcıoğlu OB, Gebauer R, Rocca D, Baroni S. turbotddft–a code for the simulation of molecular spectra using the liouville–lanczos approach to time-dependent density-functional perturbation theory. Computer Physics Communications. 2011;**182**(8): 1744-1754

[25] Timrov I. Ab initio study of plasmons and electron-phonon coupling in bismuth: From free-carrier absorption towards a new method for electron energy-loss spectroscopy [PhD thesis]. Ecole Polytechnique X; 2013

[26] Ghasemi F, Taghavimendi R, Bakhshayeshi A. Electronic and optical properties of monolayer and bulk of ptse2. Optical and Quantum Electronics. 2020;**52**(11):1-14

[27] Lei J-Q, Liu K, Huang S, Zhou X-L. The comparative study on bulk-ptse2 and 2d 1-layer-ptse2 under high pressure via first-principle calculations. Theoretical Chemistry Accounts. 2017; **136**(9):1-10

[28] Penn DR. Wave-number-dependent dielectric function of semiconductors. Physical Review. 1962; **128**(5):2093

[29] Tessier MD. Synthèse et spectroscopie optique de puits quantiques colloïdaux [PhD thesis]. Université Pierre et Marie Curie, 2013

[30] Tao W-L, Yi M, Cui-E H, Cheng Y, Ji G-F. Electronic structure, optical properties, and phonon transport in janus monolayer ptsse via first-principles study. Philosophical Magazine. 2019; **99**(8):1025-1040

[31] Asadollah Bafekry MM, Obeid CV, Nguyen MG, Bagheri Tagani M. Graphene hetero-multilayer on layered platinum mineral jacutingaite (pt 2 hgse 3): Van der Waals heterostructures with novel optoelectronic and thermoelectric performances. Journal of Materials Chemistry A. 2020;**8**(26):13248-13260

Chapter 3

Chalcogenide Materials for Sustainable Energy and Environmental Applications

Ramesh Sivasamy, Geetha Kaliyan, Selvam Kaliyamoorthy and Edgar Mosquera

Abstract

Chalcogenide materials have garnered significant attention as potential solutions for sustainable energy and environmental applications, attributed to their distinct properties and inherent advantages. This comprehensive review critically examines chalcogenide nanomaterials' emergence, synthesis, and fabrication methodologies, emphasizing their applicability across various domains, such as energy storage and conversion, photovoltaics, waste management, and water treatment processes. Furthermore, the utilization of chalcogenide nanomaterials in combating air pollution, promoting sustainable agricultural practices, and facilitating biomedical advancements is investigated. The review also encompasses an in-depth analysis of ecotoxicity and environmental safety concerns, regulatory frameworks, and policy implications for adopting chalcogenide nanomaterials. Despite the challenges and future research directions in chalcogenide nanomaterials, their global influence on sustainable development is indisputable. In conclusion, this review stresses the promising prospects of chalcogenide nanomaterials within environmental applications, underlining their potential to reshape the landscape of sustainable energy and environmental technologies.

Keywords: chalcogenides, nanomaterials, energy storage materials, photocatalysis, environmental remediation

1. Introduction

The current global energy environment is transitioning toward renewable and environmentally friendly energy sources to decrease the release of greenhouse gases and ease the consequences of climate change. To tackle these difficulties, scientific communities across the globe have been investigating innovative materials that have the potential to facilitate a more environmentally friendly, highly productive, and enduring future [1–3]. Various materials, including perovskites, skutterudite, metal-organic frameworks, covalent organic frameworks, metal chalcogenides, and intermetallic, have been recognized for their potential to generate alternative energy. Among these materials, metal chalcogenides, which comprise elements from the

IntechOpen

chalcogen group (oxygen, sulfur, selenium, and tellurium) have garnered attention because of their diverse properties. The origins of chalcogenide nanomaterials can be traced back to the initial investigation of amorphous chalcogenide semiconductors (**Figure 1a**), throughout the 1960s. Chalcogenide semiconductor nanoparticles exhibit a distinctive amalgamation of optical, electrical, thermal, and catalytic characteristics, rendering them well-suited for diverse applications in sustainable nanotechnology and environmental remediation (**Figure 1b**) [4, 5].

These characteristics encompass notable photosensitivity, substantial light absorption, and adjustable band gaps, rendering them well-suited for optoelectronics, solar cells, and photocatalysis implementation. The attractive features of these materials lie in their electronic properties, which include the ability to adjust electrical conductivity and exhibit high carrier mobility. As a result, they have garnered significant interest in their potential applications in electronic devices, such as transistors, sensors, and energy storage systems. Chalcogenide nanostructures possess distinctive thermal properties that render them well-suited for utilization in thermoelectric applications. Transition metal-based catalysts have shown remarkable efficacy in facilitating chemical reactions such as hydrogen evolution and CO_2 reduction, offering a viable and ecologically sound approach to energy production.

Chalcogenide nanomaterials possess promising prospects for environmental remediation, specifically in water purification, air pollution mitigation, and the degradation of organic contaminants [6–8]. The inherent adaptability of these systems enables the development of tailored approaches for energy conversion, electronics, and environmental remediation. The compatibility of these substances with diverse substrates and materials facilitates the production of sophisticated materials and systems that exploit synergistic features.

The chalcogenide materials listed in **Table 1** have garnered attention because of their inherent photosensitivity, rendering them well-suited for applications in data storage and optical devices. Over time, there has been a notable expansion in studies about chalcogenide nanomaterials. This expansion has resulted in substantial progress in developing fabrication techniques and a deeper comprehension of these materials' inherent features and prospective applications. This chapter aims to briefly analyze chalcogenide nanomaterials, including their distinctive characteristics and significance in sustainable energy conversion and environmental remediation.

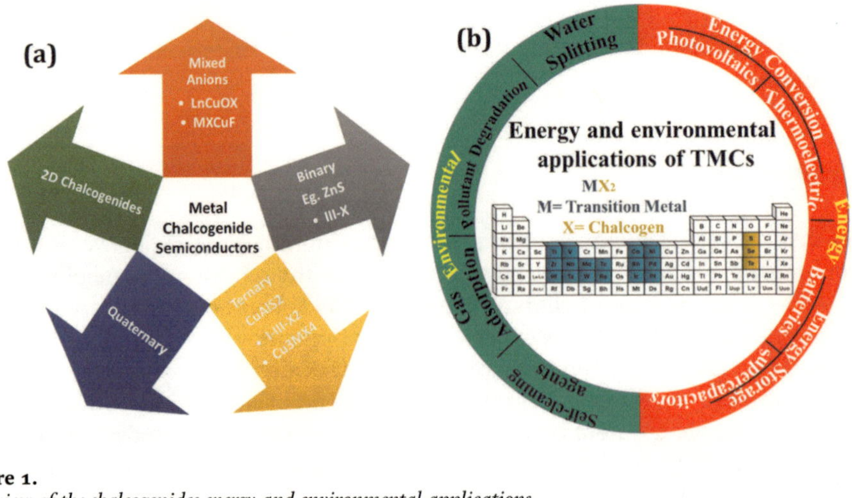

Figure 1.
Overview of the chalcogenides energy and environmental applications.

S.No	Materials	Applications	Ref.
1	CdX (X = S, Se, Te)	Solar Cells, Photocatalysis	[9]
2	Ag_2X (X = S, Se, Te)	Thermoelectric Devices, Superionic Conductors	[9–11]
3	MoS_2, WS_2	Lubricants, Catalysts, Energy Storage Devices	[11, 12]
4	Bi_2X_3, (X = S, Se, Te)	Thermoelectric Devices, Topological Insulators	[13]
5	SnS, SnSe	Solar Cells, Photocatalysis, Batteries	[14]
6	GeX (X = S, Se, Te)	Phase-Change Memory, Infrared Optics	[15]
7	Cu(In,Ga)Se2	Solar Cells	[16]
8	Cu_2ZnSnS_4	Solar Cells	[17]
9	Ti2S, FeS	Thermoelectric	[18]
10	Sb2S3, $Sb2_{(1\ x)}Fe_{2x}S_3$	Thermoelectric	[19]

Table 1.
Chalcogenide nanomaterials and their applications.

2. Metal chalcogenides for energy applications

The distinctive characteristics exhibited by these chalcogenide materials, including their notable electrical conductivity, adjustable bandgaps, and exceptional electrochemical performance, render them very suitable contenders for applications in energy conversion and systems.

2.1 Metal chalcogenides as photovoltaics

2.1.1 Introduction

Chalcogenide photovoltaic materials have emerged as a promising class of compounds for solar cell applications because of their unique properties and tunability. Solar energy has become an increasingly important renewable energy source to address the world's growing energy demands while mitigating climate change. The development of tools facilitating the efficient use of solar energy is of utmost importance. Photovoltaic (PV) solar cells are semiconductors that convert absorbed solar insolation into electrical energy. Solar cells are crucial in carbonizing energy infrastructure, reducing carbon footprints, and advancing toward a low-carbon society. Solar cells contribute significantly to these objectives besides generating clean, environmentally friendly, and sustainable power.

Materials chemists have identified several materials to observe the solar light function as PV materials. Their capacity for absorption in the visible light spectrum is restricted because of their broadband gap, which exceeds 3.2 eV. The percentage of ultraviolet-visible (UV-Vis) light that reaches the Earth's surface is 43%, and within this fraction, a mere 5% of solar energy is effectively harnessed. In contrast, around half of near-infrared (NIR) light reaches the Earth's surface, leading to a notable increase in the efficiency of solar power harvesting. In 2021, worldwide power generated from installed solar modules amounted to a mere 4.5%, while coal-based sources accounted for a far larger share of 37% [20]. Hence, it is imperative to discover novel photo-responsive materials with enhanced efficiency to facilitate more effective solar energy utilization, specifically within the UV-Vis spectrum. Developing a diverse range of new active semiconductors showing sufficient and consistent PV efficiency is a formidable challenge. There are several strategies, such as band gap engineering, enhancing surface area, and improving carrier mobility, to increase the efficiency of solar cell materials. Chalcogenide materials, which encompass a range of compounds containing elements from the chalcogenide group, offer distinct advantages in harnessing solar energy efficiently. This stimulated the scientific community to conduct thorough investigations into various PV materials and structures to attain high power conversion efficiency (PCE) while keeping costs low. Significant progress has been observed globally in recent decades, resulting in remarkable achievements. Notably, quadruple junction solar cells have attained a record-breaking PCE of 44.7% [21]. We expect that solar power will see a decrease in cost in the future years, resulting in an increase in the proportion of PV panels in the global electricity mix annually. The worldwide PV industry, valued at $100 billion, is poised for growth because of new market possibilities in various sectors, such as home, commercial, wearable, and industrial applications [22].

3. Device architecture: metal chalcogenide photovoltaics

The different layers of metal chalcogenide-based photovoltaic devices provide unique purposes that are critical to the device's overall functionality. Here, we have outlined the different layers of devices, as well as their construction and functions:

Substrate: It is a robust foundation for the device's construction. It is usually constructed of glass or similar transparent material that lets light pass through and reach the active layers.

Transparent Conductive Oxide (TCO) Layer: The TCO layer is frequently used as the front electrode and is kept on top of the substrate. It allows light into the gadget while providing electrical contact. Indium tin oxide (ITO) and fluorine-doped tin oxide (FTO) are common TCO materials.

The *window layer* is frequently utilized to increase electrical contact between the TCO and the following layers. It can also function as a diffusion barrier between layers, preventing undesired chemical reactions. It is often composed of a clear substance such as cadmium sulfide (CdS) or zinc oxide (ZnO).

Active layer: It is a device's heart, usually made of metal chalcogenide like cadmium telluride (CdTe) or copper indium gallium selenide (CIGS). This layer absorbs light energy and produces electron-hole pairs, thrusting off the photovoltaic process.

The Zn (O,S)/CdZnS might be used as a *buffer layer* and is frequently employed to improve the effectiveness of charge extraction from the absorber layer to the subsequent layers. It aids in reducing energy loss and improving overall device performance.

The Hole Passage Layer (P-type): It aids in passaging positively charged holes created in the absorber layer to the front electrode using materials such as PEDOT:PSS, spiro-OMeTAD, or CuSCN.

Electron Transport Layer (n-type): The ETL allows negatively charged electrons to go from the absorber layer to the back electrode ZnO and TiO_2 are two common ETL materials.

Cathode or Counter Electrode: The back electrode or cathode, catches the electrons carried by the ETL and completes the electrical circuit. The back electrodes can be platinum (Pt) or C-based materials.

These layers collaborate to form an effective photovoltaic device. The different layered transition metal chalcogenide-based PV solar cells are provided in **Table 2** and **Figure 2**. The precise materials and layer layouts might vary depending on the type of metal chalcogenide and the required device performance.

Different combinations of transition metal chalcogenides (TMCs), including MoS_2, CrS_2, WS_2, TiS_2, $MoSe_2$, $CrSe_2$, WSe_2, $TiSe_2$, and others, have been seen to exist in metallic, semiconductor, and insulator phases [32]. TMCs encompass a variety of large crystal families that exist in distinct phases, including 1 T, 2H, and 3R. Approximately two-thirds of these materials possess layered structures [33]. Specifically, MoS_2 has a mechanical strength that is 30% higher than steel, and it can undergo rupture after experiencing a deformation of 1%. The substance has been shown to possess exceptional distensibility and strength. Photovoltaic devices have seen a substitution of platinum (Pt) counter electrodes with molybdenum disulfide (MoS_2) in their manufacturing process [34].

Layers PV based on	Electron Transport Layer (ETL)	Hole Transport Layer (HTL)	Active Layer	Counter Electrode	Ref.
CdTe	ZnO,CdS, TiO_2	PEDOT:PSS, CuSCN, MoO3, Spiro-OMeTAD	CdTe	Pt, C, Au	[23]
CIGS	ZnO, CdS, TiO_2	PEDOT:PSS, MoO3, Spiro-OMeTAD, PTAA	CIGS	Pt, C	[24]
$CuInS_2$	ZnO, TiO_2, CdS	Spiro-OMeTAD, PTAA	$CuInS_2$		[24, 25]
MoS_2	TiO_2,ZnS, Al_2O_3	MoO3,Spiro -OMeTAD, PEDOT:PSS	MoS2		[26]
Cu_2S/CuS	ZnO,TiO_2, CdS	PEDOT: PSS, CuSCN	Cu2S/ CuS		[27]
Bi2S3	ZnO, CdS		Bi2S3		[28]
In2S3	ZnO,CdS,TiO_2		In2S3		
PbS/PbSe	TiO2, ZnO, PbSe, CuSCN	PEDOT:PSS, MoO3, Spiro-OMeTAD	PbS/ PbSe		[28]
CZTS	ZnO, CdS, TiO2, SnO2	PEDOT:PSS, CuSCN, MoO3, Spiro-OMeTAD	CuZnSnS		[28, 29]
WS2/Ag2S	TiO2, ZnO	MoO3, Spiro-OMeTAD, PEDOT:PSS	WS2/ Ag2S		[9]
FeS2/ FeSe2NiS2/ NiSe2			FeS_2/ $FeSe_2$ NiS_2/ $NiSe_2$		[30, 31]
ZnSe/ZnTe	ZnO,TiO2, CdS, MoO3	PEDOT:PSS, CuSCN, Spiro-OMeTAD	ZnSe		[5]
SnS/SnSe			SnS/SnSe		[4]

Table 2.
Presents a list of the various transition metal chalcogenides-based PV solar cells and their successive layers.

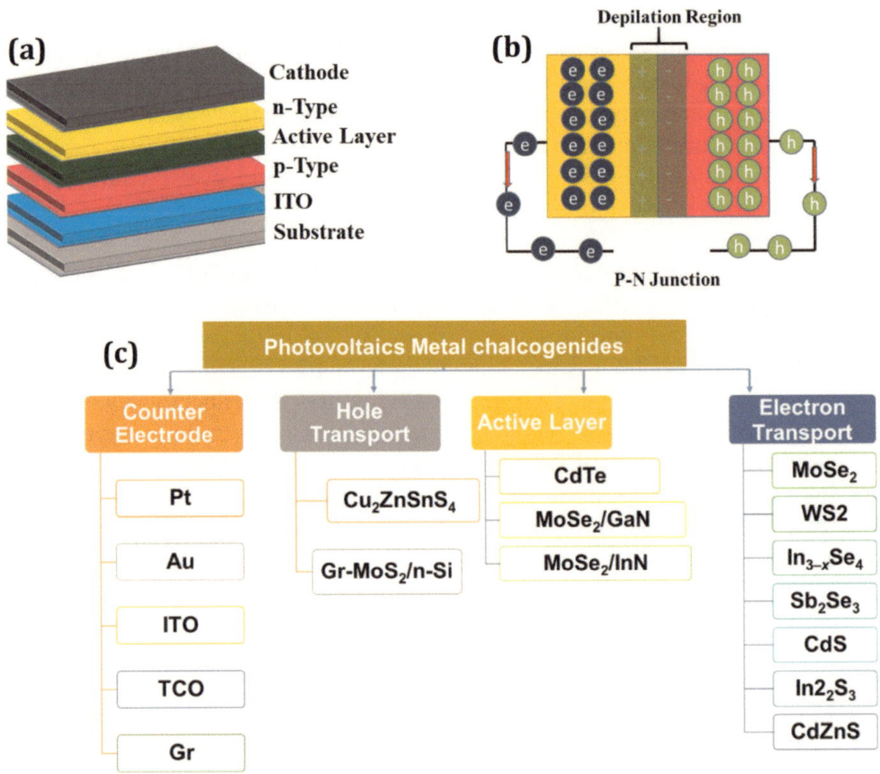

Figure 2.
(a) Schematic layout of the several layers of the solar cell, (b) P-N junction, and (c) different possible combinations of the transition metal chalcogenide layers.

4. Performance metrics for metal chalcogenide PVs

The several essential performance parameters of metal chalcogenide PV applications are below [24, 35].

- *Efficiency (η)*: It is a fundamental statistic used to quantify the effectiveness of a PV panel using metal chalcogenides in converting incoming sunlight to electrical energy. Efficiency is commonly quantified as a percentage and can be determined using the following formula: $\eta = (P_{in}/P_{out}) \times 100\%$

The variable "P_{out}" represents the electrical power output of the PV module. P_{in} represents the incident solar power on the PV module. The efficiency of metal chalcogenides can be influenced by their distinct bandgap properties, prompting endeavors to optimize this parameter.

- *Spectral response:* The spectral response of a metal chalcogenide PV cell refers to its ability to effectively capture and convert light to varying wavelengths. This aids in evaluating the efficacy of the cell over diverse lighting situations, encompassing the complete solar spectrum.

- *Short-Circuit Current (Isc):* The Isc is a parameter that denotes the highest current output achievable when the terminals of a PV cell are connected in a short-circuit configuration while adhering to conventional test parameters. The electrical properties of the metal chalcogenide material play a significant role in determining the Isc (short-circuit current) because of its absorption characteristics and electronic properties.

- *Open-circuit voltage (Voc):* Voc refers to the highest potential difference a PV cell may generate when its terminals are disconnected in a circuit with no current flow. This measurement is conducted under established test settings to ensure consistency and comparability. The determination of the metal chalcogenide's band-gap energy and other electronic properties is contingent upon several factors.

- *Fill factor (FF):* FF is a metric that quantifies the efficiency of a PV device in converting the power into electrical output. The academic definition of the term refers to the ratio between the maximum power (Pmax) and the product of the open-circuit voltage (Voc) and the short-circuit current (Isc). A more significant fill factor shows enhanced gadget performance.

- *Maximum Power Point (Pmax):* The Pmax refers to the highest level of electrical power that may be generated by a photovoltaic (PV) system. This optimal power output is achieved at the specific point when the multiplication of current and voltage is maximized. The aforementioned point is commonly referred to as the maximum power point (MPP).

- *Energy Conversion Efficiency:* The energy conversion efficiency metric evaluates a PV module's capacity to convert solar energy into electrical energy, considering solar radiation and temperature fluctuations.

- *Temperature coefficient:* It characterizes the relationship between their efficiency and changes in temperature. Comprehending these coefficients is crucial for making accurate predictions regarding real-world performance.

- *Degradation Rate:* Over time, PV devices may see a decline in performance. The degradation rate is a metric that quantifies the yearly decline in performance, offering valuable information regarding the enduring dependability of photovoltaic systems based on metal chalcogenides.

Stability and reliability measures evaluate the capacity of photovoltaic (PV) devices to sustain their performance over an extended period, encompassing their resistance to external elements, such as humidity, temperature, and ultraviolet (UV) radiation. In recent years, technical advancements in solar energy have significantly progressed, surpassing the first stages witnessed in the 19th century. However, the average efficiency of solar panels remains at approximately 15–20%. This implies that about 80–85% of the incident raw energy originating from our primary celestial body is dissipated. In addition, it is noteworthy that silicon solar cells, which represent the predominant photovoltaic technology in use, possess a theoretical efficiency limit of approximately 35% (**Table 3**).

	Single-junction			Multiple-junction		
S.No	Materials	Efficiency %	Ref.	Materials	Efficiency %	Ref.
1	Si	26.8	[36]	InGaP/GaAs/InGaAs	37.9	[37]
2	GaAs	29.1	[38]	GaInP/GaAs	32.8	[39, 40]
3	InP	24.2	[41]	GaInP/GaInAsP/S	35.9	[42]
4	CIGS	23.3	[43]	GaInP/GaAs/Si	35.9	[44]
5	CIGSSe	20.3	[45]	GaAsP/Si	23.4	[46]
6	CdTe	21.0	[47]	GaInP/GaInAs/Ge; Si	34.5	[48]
7	CZTSSe	12.1	[47]	Perovskite/Si	33.7	[48]
8	Perovskite	24.3	[49]	Perovskite/perovskite	28.2	[50]
9	GaInP	22.0	[51]	GaInP/GaAs/GaInAs	37.8	[52]
10	Organic	15.2	[53]	4 J Minimodule	41.4	[40]

Table 3.
A list of transition metal chalcogenide PVs and their efficiencies.

5. Challenges in chalcogenide photovoltaics

The materials mentioned above have exhibited exceptional performance across a range of applications. Current research and development endeavors are concentrated on several aspects to augment their effectiveness and explore untapped potential. The following are a few prominent domains in which research is currently advancing.

- *Materials Toxicity:* Certain metal chalcogenides possess hazardous elements such as cadmium and lead. Researchers are now investigating alternate materials or procedures that are less harmful to address the potential concerns of both the environment and human health. For example, the use of non-toxic substances such as copper zinc tin sulfide (CZTS) or copper zinc tin selenide (CZTSe)

- *Efficiency and Performance:* The goal of power conversion efficiencies (PCE) on par with popular technologies, such as silicon-based solar cells, has been a significant obstacle. One potential solution for achieving better efficiencies is to focus on improving the quality of crystals, optimizing processes for depositing films, and boosting the mobility of charge carriers. The investigation of band-gap engineering, tandem cell architectures, and sophisticated light trapping methods is also underway.

- *Stability and Reliability:* Certain metal chalcogenides exhibit susceptibility to environmental factors such as moisture, temperature, and light exposure, resulting in deterioration and reduced longevity. Encapsulation methods, protective coatings, and a deeper understanding of degradation processes may improve these issues.

- *Material Homogeneity and Defects:* Material defects, compositional non-uniformities, and grain boundaries may impact the transit and recombination of charges inside a material.

- *Advanced Material Engineering:* The enhancement of electrical and optical properties in metal chalcogenides may be achieved by modifying their composition and structure. This strategy holds promise for enhancing solar absorption by adjusting the band gaps by the solar spectrum.

- *Interface Engineering:* To maximize the capture of charge carriers and reduce recombination, interface engineering includes carefully designing interfaces between various materials. This promotes effective charge transport. Incorporating charge-selective layers, such as buffer layers or electron/hole transport layers, is often used in engineering to enhance the overall efficacy of devices.

- *Scalable Fabrication Methods:* The challenge is in developing economically viable and scalable methodologies for mass manufacturing metal chalcogenide solar cells on a big scale. The proposed approach involves investigating printing techniques and roll-to-roll manufacturing methods to achieve efficient and cost-effective fabrication processes.

- *Environmental Impact Reduction:* Reducing the environmental footprint of production processes by recycling and reusing materials to reduce waste. Alternative chalcogen sources and sustainable synthesis routes are investigated as a solution.

6. Emerging progress and current trends in chalcogenide photovoltaics research

Recent developments have influenced the field of solar energy conversion and changing research trends in metal chalcogenide photovoltaics. An area of significant research is the investigation of tandem and multifunction solar cells, which integrate metal chalcogenide solar cells with other photovoltaic technologies to enhance overall efficiency. Combining different elements or components enables efficient absorption of light over a broader range of solar wavelengths, resulting in improved overall performance of the device. Another area of research that is as intriguing is the exploration of hybrid architectures, which include the seamless integration of metal chalcogenides with perovskite materials. This study's primary aim is to enhance light absorption and optimize the separation of charges inside perovskite layers, with the potential to increase efficiency [54–56].

The process of bandgap engineering through anion alloying, such as in the case of $BaZrO_3$, results in the material exhibiting insulating properties with a bandgap of approximately 5 eV. In contrast, the bandgap of $BaZrS_3$ is approximately 1.8 eV. Similarly, the bandgaps of these materials can be controlled by including different anions and cations during the synthesis of chalcogenide perovskites. The substitution of titanium (Ti) on the zirconium (Zr) site and the incorporation of selenium (Se) with sulfur (S) have yielded promising outcomes. For instance, the chalcogenide perovskite $BaZr_{0.75}Ti_{0.25}S_3$ resulted in a reduced bandgap of 1.43 eV. This value is in close proximity to the optimal value, known as the Shockley-Queisser limit. In contrast, the $BaZr(S_{0.6}Se_{0.4})_3$ showed a decrease of 180 meV in the bandgap of $BaZrS_3$ [57, 58].

Selectively removing photo-generated electrons through the electron transport layer is frequent in photovoltaic devices. Through a large Schottky barrier at the perovskite material-ETL border, the electron transport layer (ETL) prevents electron vacancies from moving. The heterojunction tunneling mechanism (HTM) inhibits electron mobility, making whole extraction easier. Thus, inadequate charge transfer and charge flow obstruction will cause charge recombination at the contact, resulting in energy loss. Interface recombination involves non-radiative charge carrier recombination. Contrasting energy levels, surface imperfections, and carrier migration toward the interface cause this process.

The enhancement of understanding about defects and their subsequent impact on the functioning of devices has facilitated the development of more efficient solutions for defect passivation. By strategically implementing surface treatments and interface engineering approaches, scholars have successfully reduced recombination rates and prolonged the lifespans of charge carriers [5, 6, 8, 59–61]. Innovative techniques for characterizing materials and electronics, such as high-resolution imaging and spectroscopy, have provided significant revelations about their fundamental properties and complex operational mechanisms. Methods like Kelvin probe microscopy, terahertz spectroscopy, and time-resolved photoluminescence. Flexible and transparent substrates have facilitated the emergence of metal chalcogenide solar cells that possess flexibility and partial transparency. This technological advancement allows for smoothly incorporating these devices onto various surfaces and objects, broadening their uses.

This work investigates novel thin film device light absorption methods. These include studying plasmonic nanoparticles, nanostructured surfaces, and photonic crystals to improve light absorption. Machine learning and computer modeling speed up material discovery and device optimization. This partnership predicts material properties and determines the best experimental methods. Academics are exploring scalable fabrication methods to increase metal chalcogenide solar cell production. These methods include solution-based and roll-to-roll manufacturing.

Implementing environmentally sound synthesis processes and recycling procedures has been driven by a strong dedication to sustainability. A significant focus on replacing uncommon and dangerous elements with readily available earth-abundant resources characterized this dedication. Using metal chalcogenide solar cells in architectural components such as glass, windows, and facades produces electricity while enhancing visual appeal and promoting novel architectural concepts. The successful exceeding of the 30% criterion instills assurance that cost-effective photovoltaic systems with superior performance can be effectively introduced into the market.

Surpassing this threshold instills a sense of assurance that high-performance, cost-effective photovoltaic systems can be successfully introduced to the market. There is increasing competition among materials scientists, as seen by recent developments reported by LONGI, a prominent Chinese corporation that holds a significant market share in global solar panel production. Recently unveiled its latest innovation as a tandem solar panel with an impressive efficiency rating of 33.5% [62]. Although this situation is indeed thrilling, it is important to note that we are merely at the initial stages. Researchers have been endeavoring to address this issue for an extended period. A team at the National Renewable Energy Laboratory (NREL developed a panel with an efficiency of 47%). However, regrettably, the cost of this model renders it is impractical for widespread adoption [58]. A recent study conducted by researchers at the Fraunhofer Institute for Solar Energy Systems ISE has shown a notable enhancement in the efficiency of the most advanced four-junction solar cell to date. Implementing a novel antireflection coating achieved this improvement.

Specifically, the efficiency of the solar cell was raised from 46.1 to 47.6%. This achievement represents a significant global milestone, as there is currently no solar cell that surpasses its efficiency on a global scale [63].

6.1 Thermoelectric chalcogenides

I. *Thermoelectricity and its Importance:* The reports from the global footprint network provide an essential analysis showing a potential decline in fossil fuel reserves during the next 50 years. In recent years, the combustion of fossil fuels has had a significant impact on the ecosystem and climatic conditions. Carbon dioxide output has reached a historically high level in the last decade, a concerning indication for the future. Several data show that the energy used in the whole consumption amounts to just 40%, while the remaining energy is released into the environment as heat (**Figure 3a**). Scientific findings have bolstered the case for the immediate implementation of coordinated measures to mitigate environmental catastrophes. The increasing global energy requirements, the diminishing reserves of fossil fuels, and the growing apprehensions over climate change have prompted an extensive study of alternative energy sources and energy conversion methods. There are a few alternate techniques for mitigating the energy problem, one of which involves the recovery of spent energy wasted heat. We used thermoelectric modules as a kind of energy conversion technology [64].

The thermoelectric generator is a solid-state configuration that directly converts thermal energy into electrical energy. It can function as a solid-state refrigerator, as seen in **Figure 3(b)**. The processes mentioned above are founded upon the Seebeck and Peltier effects principles, which just need a temperature difference inside a material to generate electricity. Thermoelectric materials have several benefits, including noise reduction, compact dimensions, expandability, long-lasting performance, absence of mechanical forces, and lack of chemical reactions. The material chemist's primary goal is to find thermoelectric materials that are both efficient and environmentally beneficial while also being economically viable. The thermoelectric notion has been acknowledged for over 150 years, but its comprehension of electron and phonon transport events has solidified its foundation.

II. Thermoelectric properties: Thermoelectric figure of merit

The efficiency of converting heat to electricity in thermoelectric materials is evaluated using the dimensionless figure-of-merit, zT. This figure-of-merit is calculated using the equation $zT = S^2\sigma T/\kappa$, where S represents the Seebeck coefficient, σ represents the electrical conductivity, T represents the temperature, and κ

(a)

Energy Source	EJ (2021)	IEA-Report	
Coal	2		Energy Loss > 60%
Oil	31	E	
Natural gas	9	N	
Biomass	24	E	
Nuclear	30	R	
Hydro	16	G	
Wind	7	Y	Utilized energy < 40 %
Solar	5		
Geothermal	5		

(b)

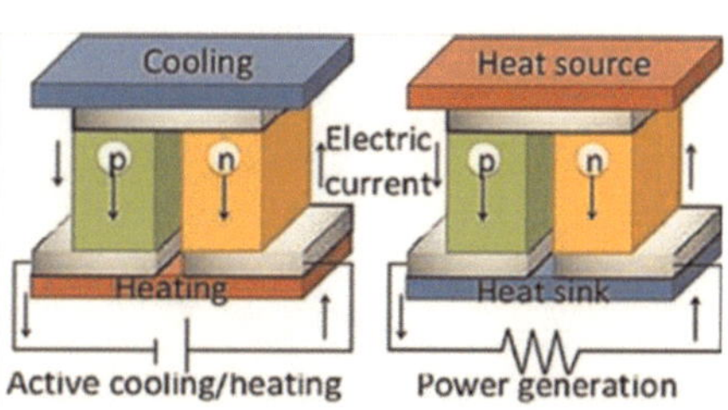

Figure 3.
Schematic representation of the Seebeck and Peltier effect.

represents the thermal conductivity. To get an adequate power output, a substantial temperature difference is imperative, denoted as $\Delta T = Th - Tc$. The good thermoelectric material must exhibit good efficiency η seen across a limited temperature difference ΔT. The highest efficiency η across a finite temperature difference ΔT. We can readily estimate this value using S, ρ, and κ.

The effectiveness of thermoelectric systems is contingent upon many parameters, including the Carnot efficiency and the thermoelectric characteristics, including the Seebeck coefficient, thermal conductivity, and electrical resistivity. The average energy conversion efficiency of about 1 ZT is 10%. If the ZT value is around 3, it results in a 20% increase. Once the thermoelectric figure of merit (ZT) reaches around 4, the conversion efficiency may increase to as high as 30%. At present, we have successfully attained a ZT value greater than 2, which has enabled its practical use. In order to attain thermoelectric materials with high efficiency, it is essential that the materials exhibit a substantial Seebeck coefficient and electrical conductivity while simultaneously maintaining a high coefficient of thermal conductivity. **Figure 4** illustrates a graphical representation of the thermoelectric module.

In contemporary times, there has been a significant focus on chalcogenides, half-Heusler compounds, clathrates, Zintl phases, and skutterudite owing to their notable high thermoelectric figure of merit (ZT). Most of the materials mentioned above are mostly composed of heavy metals and are characterized by a limited abundance. Therefore, the development of thermoelectric materials encompasses the attainment of a high figure of merit (ZT) as well as the need to operate across a broad temperature range using materials that are both non-toxic and economically viable. Most thermoelectric materials currently in the market are based on telluride compounds, with notable examples, including PbTe, Bi_2Te_3, $NaPb_xSbTe_{x+2}$, and $AgSbTe_2$. These materials offer several advantages, such as low thermal conductivity (2.3 W/m.K for PbTe and 1.7 W/m. K for Bi2Te3), a high Seebeck coefficient of 500 μV/K, and the ability to be easily doped with either p-type or n-type atoms. The exorbitant cost of tellurium-based thermoelectric materials has prompted researchers to explore alternate options. Among these alternatives, sulfur (S; Bi2S3) and selenium (Se; PbSe) have been promising candidates because of their low thermal conductivity and high Seebeck coefficients.

III. Approaches to improve the thermoelectric efficiency

a. *Doping to Modify Charge Carrier Concentration:* The doping process entails the introduction of dopants, which are foreign elements, into the lattice structure of chalcogenide materials to alter their electrical characteristics.

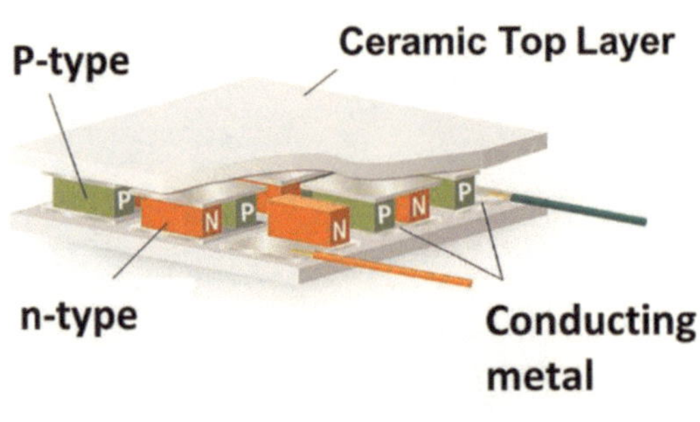

Figure 4.
Graphic representation of the thermoelectric module.

DOI: http://dx.doi.org/10.5772/intechopen.1003048

The manipulation of dopants enables the regulation of charge carrier concentration, exerting an impact on electrical conductivity and the Seebeck coefficient. In the process of n-type doping, introducing extra electrons augmented the electron concentration. Conversely, p-type doping involves the introduction of holes to elevate the engagement of holes. The selection of a dopant, together with its concentration and placement within the crystal lattice, may substantially influence the electrical characteristics of a material, affecting its thermoelectric efficiency.

b. *Nanostructuring and Grain Boundary Engineering:* The process of nanostructuring entails the deliberate manipulation of a material's structural composition at the nanoscale dimension, resulting in the formation of minute features such as nanoparticles, nanowires, or nanograins. These characteristics can disperse phonons, which are vibrations that transfer heat and increase electron scattering. As a result, the material's thermal conductivity is reduced, and its thermoelectric efficiency is improved. The primary aim of grain boundary engineering is to optimize the grain size and distribution inside a material. This is because grain borders have the potential to function as scattering foci for phonons, resulting in a decrease in thermal conductivity.

c. *Band-gap engineering:* The term "band engineering" pertains to the manipulation of the electronic band structure of a material to enhance its electronic characteristics to optimize thermoelectric performance. This process entails changing the location and breadth of the energy bands to optimize the Seebeck coefficient and electrical conductivity. The aim of energy filtering is to selectively permit the transmission of specific energy levels of charge carriers while impeding others, enhancing the material's thermoelectric characteristics. This technique shows significant use in materials characterized by intricate band topologies.

d. *Phonon Engineering and Reduction of Thermal Conductivity:* Phonons transmit thermal energy inside materials. The enhancement of efficiency in thermoelectric chalcogenides reduces thermal conductivity. Phonon engineering encompasses the deliberate introduction of scattering centers, such as nanoparticles or point defects, to impede phonons' efficient heat transfer capabilities by disrupting their propagation. In addition, using materials characterized by low phonon velocities or pronounced anharmonic might contribute to mitigating thermal conductivity.

e. *Transport Phenomena:* The efficiency of chalcogenides hinges on their ability to control thermal conductivity while maintaining high electrical conductivity. Phonon scattering plays a pivotal role in reducing thermal conductivity, involving interactions with grain boundaries, nanostructures, and point defects is equally vital, governed by electrical conductivity and the Seebeck coefficient. The engineering mentioned above exhibits interconnectivity and provides the potential for synergistic outcomes when combined.

Material	Maximum ZT	Temperature Range	Reference
PbTe	~2.2	600–900 K	[65]
Cu_2Se	~1.5	RT	[66]
Cu_2Te	~1.3		[66]
Sb_2X_3	1.1–1.8 eV		[67]
$BaZrS_3$	1.75–1.94		[68]

Table 4.
A list of transition metal chalcogenide PVs and their efficiencies.

Thermoelectric chalcogenide materials often possess complex crystal structures, ranging from layered to more intricate frameworks. Heavy elements and intrinsic anisotropy in their crystal structures contribute to their enhanced thermoelectric performance. The field of thermoelectric chalcogenides boasts several noteworthy examples that highlight their potential. One example is lead chalcogenides, specifically PbTe, known for its impressive ZT values at elevated temperatures. Additionally, copper chalcogenides (Cu_2Se and Cu_2Te) exhibit promising thermoelectric properties because of their complex crystal structures and inherent anisotropy (**Table 4**).

Challenges and Future Directions: Despite significant progress, challenges persist in enhancing the efficiency and viability of thermoelectric chalcogenides. Balancing trade-offs between electrical and thermal properties remains intricate, demanding novel material design strategies. Sustainability considerations of material sourcing and processing are vital for the widespread adoption of these materials.

The future of thermoelectric chalcogenides lies in the convergence of advanced computational modeling, materials synthesis techniques, and device engineering. Researchers can pave the way for transformative applications in energy harvesting and utilization by delving into the fundamental interactions governing their thermoelectric behavior. The versatility of thermoelectric chalcogenides finds applications in various domains. Solid-state thermoelectric generators, capable of harvesting waste heat from industrial processes or vehicle exhaust, hold promise in sustainable energy production. Their compatibility with flexible substrates paves the way for wearable energy devices, powering electronics from body heat. Innovative approaches involve integrating thermoelectric chalcogenides into hybrid systems, such as coupling them with photovoltaics for efficient solar energy conversion. Their potential in space exploration, where temperature gradients abound, showcases their utility in extreme environments.

6.2 Chalcogenide nanomaterials for energy conversion

6.2.1 Photocatalysis

The escalating environmental concerns have been driven by the expansion of the human population and the industrial revolution. The industrial sector discharged approximately 400 million metric tons of chemicals, including harmful solvents and metal ions, into the water. Besides industrial waste, many human activities, such as using pesticides, fertilizers, and the disposal of home trash, also contribute to the contamination of water bodies (see **Figure 1**). The abovementioned businesses emitted various detrimental organic substances, including formaldehyde, azo dyes, dioxins, pesticides, and heavy metals. The potential consequences of such actions threaten both human well-being and the ecological balance inside our natural water reservoirs. Each year, around 15% (equivalent to one thousand tons) of non-biodegradable textile dyes are discharged into streams and aquatic bodies through the wastewater generated by the textile industry. Typically, in the dyeing and finishing industry, an approximate range of 120–280 liters of water is utilized per kilogram of garment throughout the processing stage [69–71].

Based on an estimate provided by the World Bank, it has been determined that around 17–20% of water pollution can be attributed to the activities of textile businesses. Most organic dyes have been found to possess carcinogenic properties, posing significant health risks to both humans and marine species. These toxins might induce many health issues, including but not limited to cancer, neurological impairments, cardiovascular disorders, and digestive ailments, even when present in minimal quantities. Many actions have been identified as factors that lead to the gradual disruption of water flow in the foreseeable future (**Figure 5**) [69, 72].

Therefore, there is an urgent need to detoxify or remove deleterious chemical contaminants from water to ensure long-term sustainability. Various standard water treatment methods were accessible, such as membranes, ion exchange, filtration, remediation, coagulation, and sedimentation. Several advanced oxidation techniques, including photocatalysis, wet oxidation, and sonolysis, are recognized for their ability to mineralize resistant substances. In recent decades, there has been a growing interest in utilizing advanced oxidation processes (AOPs) to eliminate harmful organic chemicals from water [73–76]. This approach is one of several detoxification methods that has garnered significant attention. The integration of many study fields within the realm of Photocatalysis has been significantly driven by the imperative for environmental reform, clean hydrogen (H_2) fuel generation, and the conversion of CO_2 (**Figure 6**). Photocatalysis has several advantages compared to traditional catalytic processes, such as eliminating hazardous stages and operating at elevated temperatures and pressures. Studies have shown that using semiconductor photocatalysis in photo-oxidation processes yields water treatment efficiencies of over 95%.

In the water treatment process, four distinct tactics have facilitated the degradation of pollutants by Photocatalysis (**Figure 6a**): (1) Strategies involve the creation of electron-hole pairs, known as photoexcitation, which occurs when light is absorbed, (2) The process of ionization of water (H_2O), (3) the phenomenon of oxygen ionsorption, (4) the protonation of superoxide.

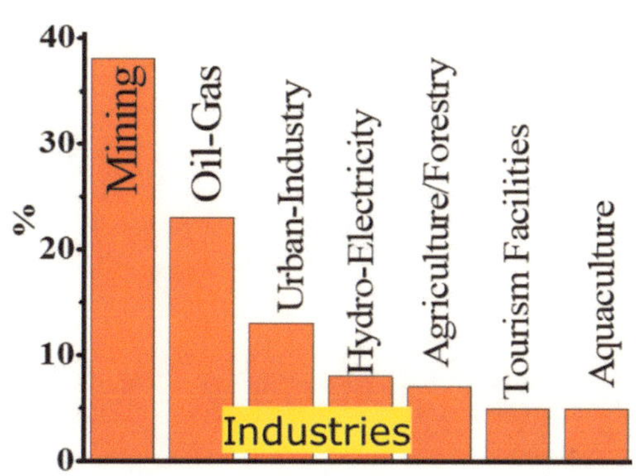

Figure 5.
Source of water pollution chemical contamination.

$$\text{Photocatalyst} + h\nu\ (\text{UV}) \rightarrow \text{Photocatalyst}\ \left(e^-(\text{CB}) + h^+(\text{VB})\right) \tag{1}$$

$$H_2O(\text{ads}) + h\nu^+(\text{VB}) \rightarrow OH^\bullet(\text{ads}) + H^+(\text{ads}) \tag{2}$$

$$O_2 + e^-(\text{CB}) \rightarrow {O_2}^{-\bullet}(\text{ads}) \tag{3}$$

$$\text{Dye} + OH^\bullet \rightarrow Co_2 + H2O\ (\text{dye intermediates}) \tag{4}$$

$$\text{Dye} + h^+(\text{VB})) \rightarrow \text{Oxidation products} \tag{5}$$

$$\text{Dye} + e^-(\text{CB}) \rightarrow \text{Reduction products} \tag{6}$$

The dynamic nature of this field causes the use of materials that are economically viable, devoid of toxicity, environmentally sustainable, competitive, and easily synthesized. Using a wide range of materials in the field has gained significant appeal. Many materials, including graphene, titanium dioxide (TiO_2), [2] zinc oxide (ZnO), copper, tin sulfide (Cu_2SnS_3), zirconium dioxide (ZrO_2), cerium dioxide (CeO_2), iron oxide (Fe_2O_3), tin dioxide (SnO_2), strontium titanate ($SrTiO_3$), tungsten trioxide (WO_3), among others, have been investigated for their potential multifunctional applications. Titanium dioxide (TiO_2) is widely acknowledged as the predominant material among semiconductors, renowned for its potential as a photocatalyst. The titanium dioxide (TiO_2) material, possessing a band-gap energy of 3.2 eV, predominantly exhibits absorption characteristics toward ultraviolet (UV) light with a wavelength of approximately 386 nanometers (nm). This specific UV wavelength accounts for a mere 5% of the total solar energy that reaches the Earth's surface. Therefore, there is a significant interest in developing a photocatalyst with a band gap lower than 3.2 eV to improve light absorption within the visible spectrum. A few photocatalytic items have been designed and commercially introduced.

In recent times, there has been a notable surge in the attention received by metal-chalcogenides among researchers. This heightened interest can be attributed to these materials' captivating electronic structure and optical capabilities, particularly in their application in photocatalysis. Tin Selenide SnS, $MoSe_2$ metal chalcogenide show a versatile material with a two-dimensional layered structure that finds utility in several fields, such as thermoelectricity, photodetection, solar energy conversion, photocatalysis, gas sensing, battery technology, and topological insulation (refer to **Figure 6b**). SnSe possesses a direct and indirect band gap ranging from 0.9 to 1.8 eV. Its high absorption coefficient, exceeding 10^4 cm^{-1}, contributes significantly to its exceptional optical absorption capabilities. Despite demonstrating significant

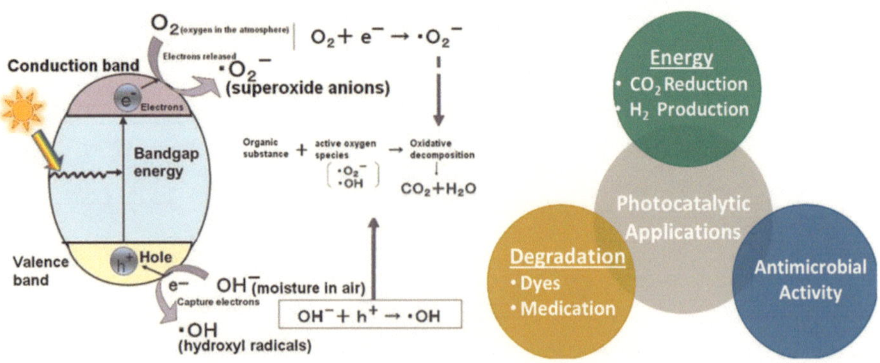

Figure 6.
Mechanism and uses of photocatalysis.

potential as a photocatalyst, further research and development are necessary because of limitations in certain aspects within a singular material composition.

One potentially helpful photocatalyst relies on three fundamental properties of semiconductor materials: 1. Band-gap, 2. Rate of separation of electron-hole pairs, and 3. Optical absorption efficiency. Within the context of a wide band-gap semiconductor, electron excitation can be observed when subjected to ultraviolet radiation, as this type of radiation possesses the energy to stimulate the electrons into an activated state. In contrast, small band-gap semiconductors can excite electrons within the visible energy range. The early excitement around this phenomenon has primarily stemmed from the properties of individual materials. However, the rapid recombination of electron-hole pairs and the limited efficiency of surface catalytic reactions pose significant challenges to the overall performance of many photocatalytic materials. Many preliminary endeavors in this domain have shown that a semiconductor photocatalyst comprising a single component cannot meet all the conditions.

We have devised various solutions to address these limitations, including using band-gap engineering via the fabrication of heterostructures. An extensive investigation in this field has shown that optimizing the energy band alignments at the interfaces of two semiconductors can significantly enhance the separation of charges by suppressing the recombination of photo-generated electron-hole pairs. Numerous endeavors have been undertaken to improve efficiency by mitigating recombination rates and broadening light absorption using surface modification, metal and non-metal ion doping, and composite formation. One promising method involves the fabrication of heterostructure nanocomposites, wherein incorporating additional semiconductors or metals has been identified as a viable strategy for augmenting their photocatalytic capabilities (see **Figure** 7) [5–8].

Multiple studies have shown that heterostructure nanosheets, interconnected through van der Waals (vdW) forces, have superior properties to individual 2D nanosheets [76, 77]. A significant number of heterostructure composites have been investigated in the field, including but not limited to $MnFe_2O_4$/rGO, $CdMoO_4$/g-C_3N_4, BiVO4/g-C_3N_4, Bi_2MoO_6/g-C_3N_4, CdxZnyS/g-C_3N_4, and Ti_3C_2/g-C_3N_4, WO_3/$ZnIn_2S_4$, CuFe2O_4/Ti_3C_2, MoS_2/TiO_2, SnO_2/ZnSe(N2H4)$_{0.5}$, $MOSe_2$/GaN, $MoSe_2$/InN,

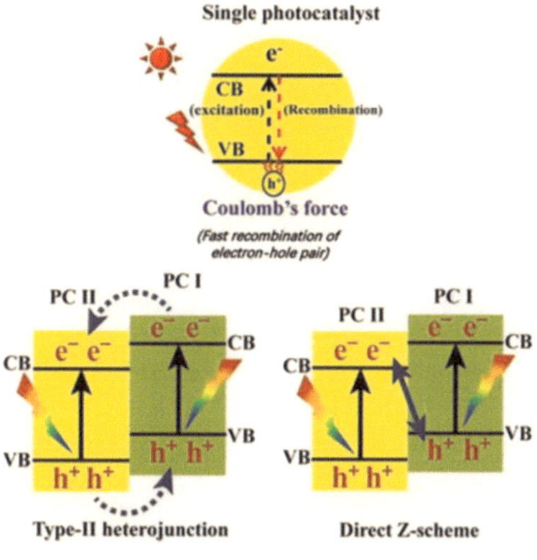

Figure 7.
Types of electron-hole migration in heterojunction photocatalyst.

and CuO/TiO_2. Besides the oxide mentioned above heterostructures, there have been reports of metal chalcogenide ($MoSe_2$, SnX, and ZnSe) heterostructures, including $SnSe_2/Se$, $SnSe/SnO_2$, $SnSe_2/SnSe$, $SnSe/g\text{-}C_3N_4$, SnO_2/ZnS, SnSe/SnO2, $MoS_2/ZnSe$, ZnTe/ZnSe, $MoSe_2/GaN$, MoSe2/InN, GaN/MoSe2/GaN, and ZnO–ZnSe. However, the number of reported metal chalcogenide heterostructures remains limited, and significant efforts are currently being made to improve their efficiency [4, 6–8, 61].

7. The future prospects of metal chalcogenides in photovoltaics, Thermoelectrics, and catalytic applications

A review of this study shows that metal chalcogenides exhibit distinct characteristics that render them promising contenders for sustainable energy generation and conversion technologies. In this concluding section, we will engage in a reflective analysis of the future prospects of these materials within the areas mentioned above while delving into their applications in these contexts.

The pursuit of highly efficient and economically viable solar energy conversion technology continues to be a primary focus in our endeavors to tackle worldwide energy-related issues. Metal chalcogenides provide many characteristics within the field of photovoltaics that present encouraging potential for the future. Metal chalcogenides have the potential to be incorporated into tandem solar cell configurations to improve the efficiency of converting solar energy. Combining materials with complementary band gaps can efficiently use the complete solar spectrum. The production of thin-film solar cells holds promise for cost reduction in manufacturing processes and the exploration of novel applications in the realm of flexible and lightweight solar panels.

Further investigation into metal chalcogenide quantum dots and nanowires might contribute to advancing innovative solar systems with high efficiency and adjustable electrical characteristics. Sustainable energy challenges include efficiently capturing and turning waste heat into electricity. Metal chalcogenides can enhance energy efficiency by harnessing waste heat in industrial operations and exhaust systems. Developing efficient, flexible, lightweight thermoelectric materials can enable portable and wearable energy harvesting systems. The non-toxic properties of many metal chalcogenides support the increasing focus on environmentally friendly materials in the energy industry. Metal chalcogenides with efficient catalytic characteristics are ideal for green and sustainable chemical manufacture and environmental remediation.

Future studies may involve applications in next-generation batteries and supercapacitors. It may also aid quantum computing, communication, and dot-based gadgets. These materials are essential to our sustainable and energy-efficient future because of their adaptability and breakthroughs in materials science and engineering. As we realize their potential, metal chalcogenides will become more relevant in twenty first-century energy and environmental challenges.

8. Conclusion

In summary, this extensive analysis shows the potential of chalcogenide materials in addressing global challenges related to sustainable energy and environmental issues. The distinctive characteristics and utilizations of these technologies encompass photovoltaics, thermoelectricity, and water treatment. It is necessary to recognize the inherent difficulties and the requirement for further investigation in the realm of chalcogenide nanomaterials. This covers the evaluation of ecotoxicity, assessment of environmental safety, and implementation of stringent regulatory measures. Chalcogenide nanomaterials are poised to have a substantial influence on global sustainable development, notwithstanding the obstacles that may arise. The utilization of chalcogenide nanoparticles has the potential to significantly transform sustainable energy and environmental solutions within a global context that prioritizes sustainability. Ongoing research and development efforts hold the potential to facilitate the emergence of a cleaner and more sustainable future by introducing promising advancements.

Author details

Ramesh Sivasamy[1*], Geetha Kaliyan[2], Selvam Kaliyamoorthy[3] and Edgar Mosquera[4]

1 Department of Sciences, School of Physical Sciences, Amrita Vishwa Vidyapeetham, Mysuru Campus, Karnataka, India

2 Department of Chemistry, SRM Institute of Science and Technology, Tamilnadu, India

3 Department of Chemistry, Mie University, Tsu, Japan

4 Departamento de Física, Universidad del Valle, Cali, Colombia

*Address all correspondence to: rameshsiva_chem@yahoo.com

References

[1] Fatimah I, Purwiandono G, Hidayat A, Sagadevan S, Kamari A. Mechanistic insight into the adsorption and photocatalytic activity of a magnetically separable γ-Fe_2O_3/ montmorillonite nanocomposite for rhodamine B removal. Chemical Physics Letters. 2022;**792**:139410

[2] Arora I, Chawla H, Chandra A, Sagadevan S, Garg S. Advances in the strategies for enhancing the photocatalytic activity of TiO_2: Conversion from UV-light active to visible-light active photocatalyst. Inorganic Chemistry Communications. 2022;**143**:109700

[3] Munusamy S, Sivasankaran RP, Sivaranjan K, Sabhapathy P, Narayanan V, Mohammad F, et al. Gallium nitride-polyaniline-polypyrrole hybrid nanocomposites as an efficient electrochemical sensor for mebendazole detection in drugs. Electrochimica Acta. 2023;**448**:142148

[4] Sivasamy R, Amirthaganesan S, Espinoza-González R, Quero F, Batoo KM. First-principles investigation of the electronic structure, optical and thermodynamic properties on monolayer Sn0.5Ge0.5Se nanosheet. Physica E: Low-dimensional Systems and Nanostructures. 2021;**126**:114454

[5] Sekar R, Sivasamy R, Ricardo B, Manidurai P. Ultrasonically synthesized TiO_2/ZnS nanocomposites to improve the efficiency of dye sensitized solar cells. Materials Science in Semiconductor Processing. 2021;**132**:105917

[6] Sivasamy R, Quero F, Paredes-Gil K, Batoo KM, Hadi M, Raslam EH. Comparison of the electronic, optical and photocatalytic properties of $MoSe_2$, InN, and $MoSe_2$/InN heterostructure nanosheet-a first-principle study. Materials Science in Semiconductor Processing. 2021;**131**:105861

[7] Sivasamy R, Paredes-Gil K, Quero F. Theoretical investigation of electronic and optical properties of the 2D-$MoSe_2$/ GaN heterostructure nanosheet. Physica E: Low-dimensional Systems and Nanostructures. 2022;**135**:114994

[8] Sivasamy R, Paredes-Gil K, Ramaclus JV, Mosquera E, Kaliyamoorthy S, Batoo KM. Sandwich-like GaN/$MoSe_2$/GaN heterostructure nanosheet: A first-principle study of the structure, electronic, optical, and thermodynamical properties. Surfaces and Interfaces. 2022;**34**:102298

[9] Kottayi R, Maurya DK, Sittaramane R, Angaiah S. Recent developments in metal chalcogenides based quantum dot sensitized solar cells. ES Energy & Environment. 2022;**18**:1-40

[10] Jin M, Lu X, Guo L, He F, Bai X, Li R, et al. Thermoelectric properties of ductile Ag2S0.7Te0.3 prepared via zone melting method. Solid State Communications. 2023;**364**:115123

[11] Zhang F, Zhu B, Guo H, Qiu J, Zheng K, Chen X, et al. Monolayer square-Ag2X (X = S, Se): Excellent n-type thermoelectric materials with high power factors. Applied Surface Science. 2021;**550**:149230

[12] Fan X, Shi Y, Cui M, Ren S, Wang H, Pu J. MoS2/WS2 Nanosheet-based composite films irradiated by atomic oxygen: Implications for lubrication in space. ACS Applied Nano Materials. 2021;**4**:10307-10320

[13] Guo M, Wang Z, Xu Y, Huang H, Zang Y, Liu C, et al. Tuning

thermoelectricity in a Bi_2Se_3 topological insulator via varied film thickness. New Journal of Physics. 2016;**18**:015008

[14] Shen Y, Zhang Y, Huo J, Li X, Yan Z, Pan Y, et al. Two-dimensional SnSe material for solar cells and rechargeable batteries. Journal of Energy Storage. 2023;**69**:107958

[15] Cao T, Wang R, Simpson RE, Li G. Photonic Ge-Sb-Te phase change metamaterials and their applications. Progress in Quantum Electronics. 2020; **74**:100299

[16] Barde A, Lewis DJ. Fabrication of high quality Bornite and chalcopyrite thin films by aerosol-assisted chemical vapor deposition. The Journal of Physical Chemistry C. 2023;**127**:13969-13977

[17] Yan C, Huang J, Sun K, Johnston S, Zhang Y, Sun H, et al. Cu2ZnSnS4 solar cells with over 10% power conversion efficiency enabled by heterojunction heat treatment. Nature Energy. 2018;**3**: 764-772

[18] James ACWP, Goodenough JB, Clayden NJ, Banks PM. Characterization of defect thiospinels Cu1−x[Ti2]S4 (0<x≤0.93). Materials Research Bulletin. 1989;**24**:143-155

[19] Almalki R, Mkawi EM, Al-Hadeethi Y. Fabricating antimony sulfide Sb2S3 microbars using solvothermal synthesis: Effect of the solvents used on the optical, structural, and morphological properties. Journal of Materials Science: Materials in Electronics. 2020;**31**:9203-9211

[20] World Energy Balances Overview, International Energy Agency, Paris. 2021. Available from: https://www.iea.org/reports/world-energy-balances-overview. License: CC BY 4.0

[21] Alharbi FH, Kais S. Theoretical limits of photovoltaics efficiency and possible improvements by intuitive approaches learned from photosynthesis and quantum coherence. Renewable and Sustainable Energy Reviews. 2015;**43**: 1073-1089

[22] Sasikala R, Kandasamy M, Ragavendran V, Suresh S, Sasirekha V, Murugesan S, et al. Perovskite zinc titanate-reduced graphene oxide nanocomposite photoanode for improved photovoltaic performance in dye-sensitized solar cell. Physica B: Condensed Matter. 2022;**646**:414300

[23] Loh L, Dunn S. Recent progress in ZnO-based nanostructured ceramics in solar cell applications. Journal of Nanoscience and Nanotechnology. 2012; **12**:8215-8230

[24] Miah MH, Rahman MB, Nur-E-Alam M, Das N, Soin NB, Hatta SFWM, et al. Understanding the degradation factors, mechanism and initiatives for highly efficient perovskite. Solar Cells. 2023;**9**:e202200471

[25] Ašmontas S, Mujahid M, Recent progress in perovskite tandem solar cells. Nanomaterials 2023;**13**:1886

[26] Zhu X, Zhang R, Li M, Gao X, Zheng C, Chen R, et al. PEDOT:PSS/CuCl composite hole transporting layer for enhancing the performance of 2D Ruddlesden-popper perovskite solar cells. The Journal of Physical Chemistry Letters. 2022;**13**:6101-6109

[27] Xu Y-F, Wu W-Q, Rao H-S, Chen H-Y, Kuang D-B, Su C-Y. CdS/CdSe co-sensitized TiO_2 nanowire-coated hollow spheres exceeding 6% photovoltaic performance. Nano Energy. 2015;**11**: 621-630

[28] Wang G-Q, Ren X-X, Wei J-J, Wang A-J, Zhao T, Feng J-J, et al. Ultrasensitive PEC cytosensor for breast cancer cells

detection and inhibitor screening based on plum-branched CdS/Bi2S3 heterostructures. Bioelectrochemistry. 2023;**152**:108442

[29] Njema GG, Kibet JK. A review of the technological advances in the design of highly efficient perovskite solar cells. International Journal of Photoenergy. 2023;**2023**:3801813

[30] Livingston LMM, Raj AGS, Prabu RT, Kumar A. Computational analysis of FeS_2 material for solar cell application. Optical and Quantum Electronics. 2023;**55**:244

[31] Razamin NAY, Woo HJ, Winie T. Comparative study of nickel selenide, iron selenide and platinum on triiodide reduction for dye-sensitized solar cells. Optical Materials: X. 2022;**13**: 100119

[32] Wilson JA, Yoffe AD. The transition metal dichalcogenides discussion and interpretation of the observed optical, electrical and structural properties. Advances in Physics. 1969;**18**: 193-335

[33] Velický M, Toth PS. From two-dimensional materials to their heterostructures: An electrochemist's perspective, applied. Materials Today. 2017;**8**:68-103

[34] Iqbal MZ, Alam S, Faisal MM, Khan S. Recent advancement in the performance of solar cells by incorporating transition metal dichalcogenides as counter electrode and photoabsorber. International Journal of Energy Research. 2019;**43**:3058-3079

[35] Celik I, Hosseinian Ahangharnejhad R, Song Z, Heben M, Apul D. Emerging photovoltaic (PV) materials for a low carbon economy. Energies (Switzerland). 2020;**13**(1-10):4131

[36] Genshun W, Shi Y, Fuguo P, Chengjian H, Minghao Q, Junxiong L, et al. Silicon heterojunction solar cells with up to 26.81% efficiency achieved by electrically optimized nanocrystalline-silicon hole contact layers. Nature Energy. 2023;**13**:789-799

[37] Yamaguchi M, Takamoto T, Araki K, Ekins-Daukes N. Multi-junction III–V solar cells: Current status and future potential. Solar Energy. 2005;**79**:78-85

[38] Mattos LS, Scully SR, Syfu M, Olson E, Yang L, Ling C, et al. New module efficiency record: 23.5% under sun illumination using thin-film single-junction GaAs solar cells. In: 2012 38th IEEE Photovoltaic Specialists Conference. Austin, TX, USA. 2012. pp. 003187-003190

[39] Verma M, Mishra GP. Analytical model of InP QWs for efficiency improvement in GaInP/Si dual junction solar cell. Physica Status Solidi A. 2023; **220**:2200500

[40] Zhu L, Wang Y, Pan X, Akiyama H. Theoretical modeling and ultra-thin design for multi-junction solar cells with a light-trapping front surface and its application to InGaP/GaAs/InGaAs 3-junction. Optics Express. 2022;**30**: 35202-35218

[41] Raj V, dos Santos TS, Rougieux F, Vora K, Lysevych M, Fu L, et al. Indium phosphide based solar cell using ultra-thin ZnO as an electron selective layer. Journal of Physics D: Applied Physics. 2018;**51**:395301

[42] Dimroth F, Grave M, Beutel P, Fiedeler U, Karcher C, Tibbits TND, et al. Wafer bonded four-junction GaInP/GaAs//GaInAsP/GaInAs concentrator solar cells with 44.7% efficiency. Progress in Photovoltaics:

Research and Applications. 2014;**22**: 277-282

[43] Ahmadpanah FS, Orouji AA, Gharibshahian I. Improving the efficiency of CIGS solar cells using an optimized p-type CZTSSe electron reflector layer. Journal of Materials Science: Materials in Electronics. 2021; **32**:22535-22547

[44] Feifel M, Lackner D, Ohlmann J, Benick J, Hermle M, Dimroth F. Direct growth of a GaInP/GaAs/Si triple junction solar cell with 22.3% AM1.5g efficiency. Solar Rapid Research Letter. 2019;**3**:1900313

[45] McLeod SM, Hages CJ, Carter NJ, Agrawal R. Synthesis and characterization of 15% efficient CIGSSe solar cells from nanoparticle inks. Journal of Nanoparticle Research. 2015; **23**:1550-1556

[46] Lepkowski DL, Grassman TJ, Boyer JT, Chmielewski DJ, Yi C, Juhl MK, et al. 23.4% monolithic epitaxial GaAsP/Si tandem solar cells and quantification of losses from threading dislocations. Solar Energy Materials and Solar Cells. 2021;**230**:111299

[47] Green MA, Dunlop ED, Yoshita M, Kopidakis N, Bothe K, Siefer G, et al. Solar cell efficiency tables (version 62). Progress in Photovoltaics. 2023;**31**: 651-663

[48] García I, Barrutia L, Dadgostar S, Hinojosa M, Johnson A, Rey-Stolle I. Thinned GaInP/GaInAs/Ge solar cells grown with reduced cracking on Ge|Si virtual substrates. Solar Energy Materials and Solar Cells. 2021;**225**:111034

[49] Shu H, Peng C, Chen Q, Huang Z, Deng C, Luo W, et al. Strategy of enhancing built-in field to promote the application of C-TiO_2/SnO_2 bilayer electron transport layer in high efficiency perovskite solar cells (24.3%). Small. 2022;**18**:2204446

[50] Yao M, Marzouki A, Hao S, Salmanov S, Otonicar M, Loyau V, et al. Grain size and piezoelectric effect on magnetoelectric coupling in BFO/PZT perovskite-perovskite composites. Journal of Alloys and Compounds. 2023; **948**:169731

[51] Shoji Y, Oshima R, Makita K, Ubukata A, Sugaya T. 1.5eV GaInAsP solar cells grown via hydride vapor phase epitaxy for low-cost GaInP GaInAsP//Si triple-junction structures. Advanced Energy Sustainability Resesrch. 2023;**4**: 2200198

[52] Mehrotra A, Freundlich A. Superior radiation and dislocation tolerance of IMM space solar cells. In: 2012 38th IEEE Photovoltaic Specialists Conference. Austin, TX, USA. 2012. pp. 003150-003154

[53] Li M, Wang J, Jiang A, Xia D, Du X, Dong Y, et al. Metal organic framework doped Spiro-OMeTAD with increased conductivity for improving perovskite solar cell performance. Solar Energy. 2019;**188**:380-385

[54] Guo Z, Jena AK, Kim GM, Miyasaka T. The high open-circuit voltage of perovskite solar cells: A review. Energy & Environmental Science. 2022;**15**:3171-3222

[55] Werlinger F, Segura C, Martínez J, Osorio-Roman I, Jara D, Yoon SJ, et al. Current progress of efficient active layers for organic, chalcogenide and perovskite-based solar cells: A perspective. Energies. 2023;**16**:5868

[56] Matta SK, Tang C, O'Mullane AP, Du A, Russo SP. Density functional theory study of two-dimensional post-

transition metal chalcogenides and halides for interfacial charge transport in perovskite solar cells. ACS Applied Nano Materials. 2022;**5**:14456-14463

[57] Isikgor FH, Zhumagali S, Merino LVT, De Bastiani M, McCulloch I, De Wolf S. Molecular engineering of contact interfaces for high-performance perovskite solar cells. Nature Reviews Materials. 2023;**8**:89-108

[58] Devendra T, Oliver SH, Giulia L. Chalcogenide perovskites for photovoltaics: current status and prospects. Journal of Physics: Energy. 2021;**3**:034010 (1-9)

[59] Lavín A, Sivasamy R, Mosquera E, Morel MJ. High proportion ZnO/CuO nanocomposites: Synthesis, structural and optical properties, and their photocatalytic behavior. Surfaces and Interfaces. 2019;**17**:100367

[60] Ramesh S, Marutheeswaran S, Ramaclus JV, Paul DC. Electronic structure study on 2D hydrogenated Icosagens nitride nanosheets. Superlattices and Microstructures. 2014; **76**:213-220

[61] Sivasamy R, Srinivasan M, Espinoza-González R, Mosquera E. Electronic and optical studies on two-dimensional hydrogenated stirrup triels nitride nanosheets: A first-principle investigation. Materials Science and Engineering: B. 2021;**264**:114978

[62] Akhil S, Akash S, Pasha A, Kulkarni B, Jalalah M, Alsaiari M, et al. Review on perovskite silicon tandem solar cells: Status and prospects 2T, 3T and 4T for real world conditions. Materials and Design. 2021;**211**:110138

[63] Dimrot F. World's Most Efficient Solar Cell with 47.6 Percent Efficiency. Press release Fraunhofer Institute for Solar Energy Systems ISE. 2022:1-3

[64] Lee J-H, Kim J, Kim TY, Al Hossain MS, Kim S-W, Kim JH. All-in-one energy harvesting and storage devices. Journal of Materials Chemistry A. 2016;**4**:7983-7999

[65] Khshanovska O, Parashchuk T, Horichok I. Estimating the upper limit of the thermoelectric figure of merit in n- and p-type PbTe. Materials Science in Semiconductor Processing. 2023;**160**: 107428

[66] Xing C, Zhang Y, Xiao K, Han X, Liu Y, Nan B, et al. Thermoelectric performance of surface-engineered Cu1.5–xTe–Cu2Se nanocomposites. ACS Nano. 2023;**17**:8442-8452

[67] Barthwal S, Kumar R, Pathak S. Present status and future perspective of antimony chalcogenide (Sb2X3) photovoltaics. ACS Applied Energy Materials. 2022;**5**:6545-6585

[68] Sopiha KV, Comparotto C, Márquez JA, Scragg JJS. Chalcogenide perovskites: Tantalizing prospects. Challenging Materials. 2022;**10**:2101704

[69] Sibhatu AK, Weldegebrieal GK, Sagadevan S, Tran NN, Hessel V. Photocatalytic activity of CuO nanoparticles for organic and inorganic pollutants removal in wastewater remediation. Chemosphere. 2022;**300**: 134623

[70] Koteeswari P, Sagadevan S, Fatimah I, Kassegn Sibhatu A, Razak SIA, Leonard E, et al. Green synthesis and characterization of copper oxide nanoparticles and their photocatalytic activity. Inorganic Chemistry Communications. 2022;**144**: 109851

[71] Lett JA, Alshahateet SF, Fatimah I, Sivasankaran RP, Sibhatu AK, Le M-V, et al. Hydrothermal synthesis and photocatalytic activity of Mn_3O4 nanoparticles. Topics in Catalysis. 2023; **66**:126-138

[72] Abel Noelson E, Anandkumar M, Marikkannan M, Ragavendran V, Thorgersen A, Sagadevan S, et al. Excellent photocatalytic activity of Ag_2O loaded ZnO/NiO nanocomposites in sunlight and their biological applications. Chemical Physics Letters. 2022;**796**: 139566

[73] Sagadevan S, Alshahateet SF, Anita Lett J, Fatimah I, Poonchi Sivasankaran R, Kassegn Sibhatu A, et al. Highly efficient photocatalytic degradation of methylene blue dye over Ag_2O nanoparticles under solar light irradiation. Inorganic Chemistry Communications. 2023;**148**:110288

[74] Krishna Veni K, Agalya S, Kavitha R, Fatimah I, Sagadevan S, Nehru LC. Efficient photocatalytic activity of chitosan/tin oxide nanocomposite for environmental remediation. Inorganic Chemistry Communications. 2023;**154**: 110865

[75] Fathima Khyrun SM, Jegatha Christy A, Mayandi J, Sagadevan S. Photo-triggered antibacterial and catalytic activities of solution combustion synthesized CeO_2/NiO binary nanocomposite. Inorganic Chemistry Communications. 2023;**153**: 110860

[76] Qutub N, Singh P, Sabir S, Sagadevan S, Oh W-C. Enhanced photocatalytic degradation of acid blue dye using CdS/TiO_2 nanocomposite. Scientific Reports. 2022;**12**:5759

[77] Qutub N, Singh P, Sabir S, Umar K, Sagadevan S, Oh W-C. Synthesis of polyaniline supported CdS/CdSZnS/CdS-TiO_2 nanocomposite for efficient photocatalytic applications. 2022;**12**:1355

Chapter 4

Chalcogenides-Based Nanomaterials for Contaminant Removal in Wastewater Treatment

Arunkumar Priya and Suresh Sagadevan

Abstract

The pollution has been increasing day by day which highly affects the environment. The longer we wait to take action to save the environment, the harder it will be. Increasing organic and inorganic waste production has made widespread pollution and water contamination due to rapid growth in population. It is believed that contaminated water poses a significant danger to water security. Precipitation, adsorption, electrochemical, photocatalysis, and membrane filtration are just some of the methods for purifying the water supply. One of the most efficient methods for eliminating dissolved metal ions from wastewater is photocatalysis. High efficiency, cost-effectiveness, avoiding residual pollutants, and direct application of solar energy are only a few of the benefits of the photocatalytic approach compared to other methods. Due to their lower band gaps, charge carrier mobility, and visible-light absorption, nanomaterials based on chalcogenides are widely employed as photocatalysts. A more significant number of active sites per unit surface area and a longer distance over which charge carriers could diffuse are two novel qualities that emerged due to the quantum size effect, caused by the reduction in the size of chalcogenides. In this chapter, we will dive deep into the novel application of nanomaterials based on chalcogenides for contaminant removal in wastewater treatment. Water contamination, its treatment, and the other environmental toxins are explored in detail. These chalcogenide-based compounds are used as catalysts to purify water from industrial wastes and remove environmental toxins.

Keywords: chalcogenides, nanomaterials, water treatment, transition metal chalcogens, composites, water treatment processes

1. Introduction

The current economic status depends on industrialization, making it one of the most sought-after requirements in all areas. The rate at which pollutants are discharged into the environment, and thus the severity of pollution from all sources, rises in tandem with the rate of industrialization [1]. Humans, animals, and other species dependent on water for survival are all negatively impacted by water pollution because of the widespread use of dangerous chemicals dumped into water sources. It is estimated that 2.2 million people worldwide die yearly from drinking contaminated water in the developed and developing countries. Organic and inorganic contaminants can be found in wastewater [2]. Emerging contaminants include pharmaceuticals, endocrine disruptors, personal care products, other refractory organic pollutants, and more common pollutants like dyes and heavy metals. As such, water pollution is a multifaceted environmental issue that calls for a comprehensive approach.

Chlorination is an inefficient water treatment process for removing inorganic contaminants, organic molecules, heavy metals, and more complex compounds. When substances are introduced into water and subsequently released into rivers, it poses a threat to the entire water purification system. Common issues are typically caused by an imbalance between the chemical oxygen demand and the biological oxygen demand of the aquatic organism [3]. Heavy metal ions (HMIs), azo dyes, and other organic contaminants are the most common types of potentially toxic elements (PTEs) in wastewater [4–6]. Waterborne disease epidemics affects the lives of millions of people every year, primarily in developing nations where clean water is scarce [7].

Humans are particularly susceptible to the health dangers caused by HMI, including Pb, Hg, and Cd, even at relatively modest exposure levels. Fatigue, thirst, depression, headaches, dementia, gene mutation, Wilson disease, kidney or liver damage, neurological dysfunctions, cancer, sleeplessness, diarrhea, and mortality are only some of the outcomes [8]. Human health and economic development depend on effective methods for cleaning polluted water supplies for consumption and other uses. Toxins in wastewater have been effectively eliminated using various methods including electrochemical processes, absorption, ultraviolet radiation, chemical oxidation, and different, complex filtration techniques [9]. Microorganisms' contribution to wastewater treatment, alongside more conventional approaches, are significant. Various enzymatically driven organisms, such as metal oxidation/reduction and even precipitation via enzymatic ligand synthesis, are involved in treating wastewater by microbes [10]. It is generally accepted that nanomaterials aid in developing cutting-edge methods for treating wastewater. Nanotechnology research leads to the development of nanomaterials, which are the results of nanoscale engineering. The term "nanomaterial" refers to a broad category of substances. Material properties are process- and structure-dependent. Nanostructured materials are distinct from bulk materials in ways beyond size alone [11]. These chemical reactivity, energy absorption, and biological mobility differ from those of bulk materials. These materials have recently entered the realm of cutting-edge healthcare. From the agents to targeted gene transfer, they have found several uses in medical imaging [12]. These buildings, however, raise several ecological and societal concerns. Materials like sulfides, selenites, and tellurites are some of the examples of nanostructured metal chalcogenides (MCs) because they include at least one metal cation and a chalcogen.

Materials and nanoparticles based on metal chalcogenides are widely used in energy conversion and storage devices like fuel cells, solar cells, light-emitting diodes (LEDs), ion batteries, supercapacitors, thermoelectric devices, semiconductor diode lasers, photovoltaic cells, optoelectronics, sensors, and bioelectronic components due to the remarkable physical-chemical properties of the compounds formed during the combination [13]. Hydrogen release from water splitting, wastewater treatment, and selective oxidation are just a few applications of chalcogenides as photocatalysts [14]. Nanotechnology has resulted in a proliferation of approaches for cleaning up wastewater. Current technologies that make use of nanoscale chalcogenides are discussed in this chapter. Some examples include using chalcogenide nanomaterials as catalysts for the disinfection of wastewater bacterial communities and the breakdown of organic pollutants in polluted water treatment.

2. Basics of chalcogenides and properties

Chalcogenides are a class of compounds with a wide range of structural complexity and exciting chemical and physical properties. Chalcogenides has been ideal for visible-light harvesting and related applications because the band gaps between their energies are lower than those of most oxides. Among chalcogenides, metal sulfides have drawn the most significant number of studies because of their low band gap energy, better light harvesting, and diverse applications [15]. It is possible to classify chalcogenides into three distinct categories based on the presence of either alkali metals or alkaline-earth metals, transition metals, or chemicals from the primary group. Therefore, this chapter of the book explores the various chalcogenide classification systems. Photocatalytic activity of metal sulfides, selenides, and tellurides, as well as general chalcogenide properties, are also covered [16]. Some of the characteristics that set chalcogenides apart are listed below.

- All chalcogens react strongly with alkali earth metals.
- The ionic forms of all chalcogens can be found in metallic ores.
- The less dense chalcogens vital to all life forms are not harmful.
- Tellurium, selenium, and polonium are examples of heavier chalcogens that are poisonous and potentially dangerous.
- The atomic sizes of chalcogens exhibit significant variation. Nevertheless, these elements possess a common characteristic: they each possess six valence electrons.
- A higher atomic weight results in a greater density, higher melting and boiling temperatures, and a larger nuclear radius.
- Nonmetals including oxygen, sulfur, and selenium and semimetals like tellurium and polonium conduct electricity.

When a chalcogenide bonds to a transition metal, two different structures can result: dichalcogenides and tetrahedral structures. This system will form if chalcogen is bonded to specific transition metals. When chalcogenide elements connect with Mn, Co, Fe, Cu, Ni, and Zn, a tetrahedral metal-chalcogenide structure is created, but a dichalcogenide structure is generated when Ti, Cr, V, Zr, Mo, Nb, Tc, Ta, Hf, W, and Re are present [17]. Due to their unique properties, chalcogenides are an effective solid-state electrolyte for batteries. Because of the wide variety of metal halides and chalcogenides, researchers have split them into three distinct classes based on their physical and chemical characteristics: the primary category, the phase transition group, and the group of rare earth metal chalcogenide halides [18]. The properties of several chalcogenides are shown in **Table 1**. The classification of chalcogenides is shown in **Figure 1**.

2.1 Chalcogenide classification based on the number of constituents

Binary, ternary, and quaternary systems, the presence or absence of metals, the presence or absence of chalcogen ions, and so on are all used to categorize chalcogenides.

Chalcogenide type	Chalcogenide property
Chalcogenides of alkali metals and alkaline-earth elements	Compounds that are salt-like and typically colorless and soluble in water.
Chalcogenides of transition metals	Displaying a covalent, nonionic nature, tetrahedral bonding, the crystal has a hexagonal symmetry.
Chalcogenides with a nanostructure	Facilitate the segregation of electrons and phonons with their exceptional qualities; materials have a cube-like (or rock crystal-like) form.
Glasses made from chalcogenide elements	Act like semiconductors and display the characteristics of amorphous semiconductors with band gap energies between 1 and 3 eV. So-called "lone pair" semiconductors, they have features in between those of organic polymers and oxide glasses. Unlike organic polymers, the structure of atomic bonds is relatively stable. They exhibit more pliability than oxide glasses.

Table 1.
Properties of chalcogenides [19].

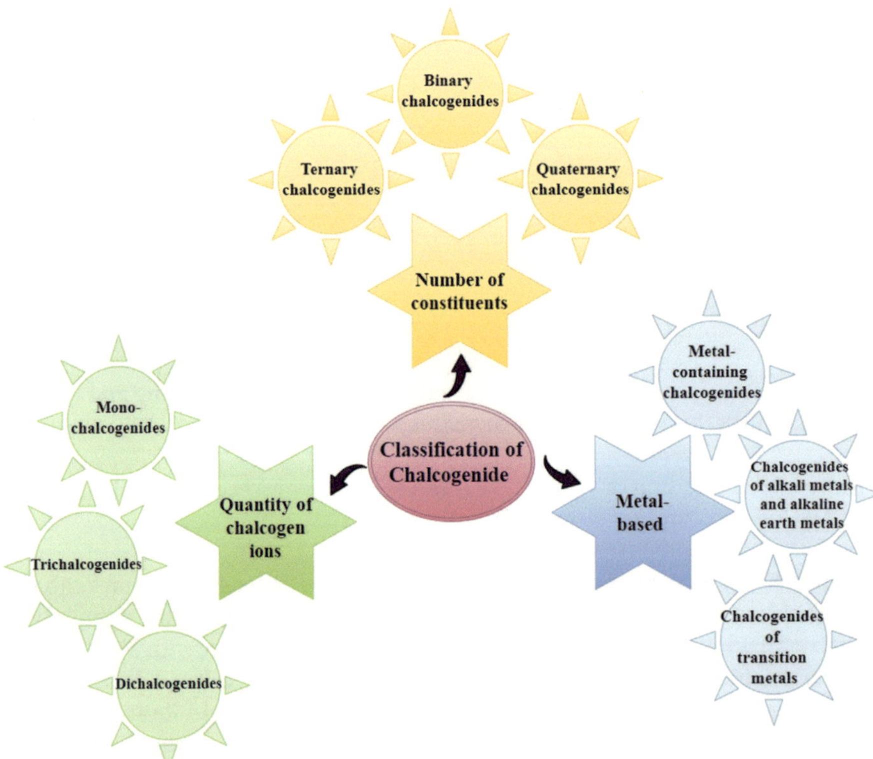

Figure 1.
Classification of chalcogenides.

In contrast to quaternary chalcogenides, extensive research has been conducted on binary and ternary chalcogenide molecules. Chalcogen derivatives exist for every element in the periodic table, except the noble gases. Their stoichiometries, such as silicon disulfide (SiS_2), boron sulfide (B_2S_3), and antimony trisulfide (Sb_2S_3), typically adhere to the common valence trends [20]. Directional covalent bonding, rather than close packing, determines the architecture of many primary group materials.

2.1.1 Binary chalcogenides

Metal cations and a chalcogen anion make the ions in binary chalcogenides. One of the most researched chalcogenides is cadmium sulfide (CdS). It is relatively active when exposed to visible light and has a band gap energy of 2.3 eV. Its unusual optical and electrical properties, which vary in size and structure, make it useful in various applications. Due to their wide range of possible uses, cadmium chalcogenides are the most essential materials, particularly for photoluminescence devices and solar cells. It was determined by Bajpai et al. that cadmium thiolate powder, when heated, nucleates to CdS particles of the desired size [21].

2.1.2 Ternary chalcogenides

Adding a third group III, IV, or V element to a binary chalcogenide produces a ternary chalcogenide. Adding a group III or IV metal to a binary chalcogenide, such as copper sulfide (CuS), results in the formation of a ternary chalcogenide. Group VI of the periodic table contains the elements sulfur and selenium. They have many properties and can be found in hydrocarbon compounds in various mixtures [22]. In addition, sulfur and selenium are frequently employed together or separately in ternary and quaternary chalcogenide compounds. There has been three ions in ternary chalcogenides; the chalcogen anion is one of them. The ability to precisely control electronic characteristics by adjusting the Cd:Zn atomic ratio makes ternary chalcogenides like ZnCdX (where X = S, Se) particularly significant. The band gap energy of zinc cadmium sulfide (ZnCdS), for instance, can range from about 2.4–3.7 eV. ZnCdS has many applications, including solar cells with heterojunctions, cathodoluminescence at low voltages, UV diodes, etc. Liu et al. reported the thermochemical synthesis of one-dimensional (1D) nanostructures of ternary chalcogenide from ZnS and CdS precursors. As-prepared 1D nanoscale ZnCdS nanotubes have shown promise in emerging nanoelectronics and photonics applications [20].

2.1.3 Quaternary chalcogenides

In addition, the quaternary chalcogenide formed by substituting half the indium in copper indium sulfide ($CuInS_2$) with zinc and another half with tin is a derivative compound from chalcopyrite, or ternary chalcogenide, consisting of the I2-II-IV-VI4 quaternary semiconductor compound. The kesterite or stannite structure of $CuZnSnS_4$ (CZTS (copper, zinc, tin, and sulfide))/Se_4 results from this substitution. The only difference between the two structures is their arrangement Cu and Zn atoms. Because of its excellent thermodynamic stability, CZTS material typically occurs in the kesterite phase rather than the stannite form [23]. A chalcogen anion is one of the four ions present in quaternary chalcogenides. Quaternary chalcogenides are a large class of materials that have several potential applications. These include serving as absorbers in solar cells, photocatalysts for solar water splitting, etc. Unlike in binary or ternary chalcogenide, many elements can be accommodated in quaternary materials, allowing for relatively complex electrical and structural features. It has been reported that the visible-light absorption of a series of sulfides quaternary chalcogenides, including $AgInZn_7S_9$, is superior to that of their oxide counterparts. Therefore, when exposed to visible light, these sulfides exhibit high photocatalytic activity for the generation of H_2 [24].

2.2 Metal-based chalcogenide classification

Different varieties of chalcogenides can be distinguished by the amount and types of metals they contain.

2.2.1 Chalcogenides of alkali metals and alkaline-earth metals

Common examples of mono-chalcogenides include alkali metals and alkaline-earth metals. Hydrolysis of mono-chalcogenide sulfides of alkali and alkaline-earth metals results in the formation of derivatives containing disulfide anions. They are chemical substances that resemble salts but lack color and generally soluble in water. The anti-fluorite structure and the sodium chloride motif are frequently observed during the crystallization of alkali metal chalcogenides. KLiX (X = S, Se, Te) is an example of an alkali metal chalcogenide, while MX (M = Ca, Sr., Ba, and X = S, Se, Te) is an example of an alkaline-earth metal chalcogenide [24, 25].

2.2.2 Chalcogenides of transition metals

Many different stoichiometries and structural forms of transition metal chalcogenides (TMCs) have been discovered. Commonly studied transition metal chalcogenides include zinc (ZnS, ZnSe, and ZnTe), cadmium (CdS, CdSe, and CdTe), molybdenum (MoS2), and mercury (HgCdTe). TMCs, with their zinc-blend or wurtzite structure, exhibit strongly covalent, nonionic activity. TMCs can be broken down, even further by subcategories [21, 25].

2.2.3 Metal-containing chalcogenides

In metal-rich chalcogenides, transition metals exhibit II or higher oxidation states. These compounds usually include necessary metal-to-metal bonding and can be distinguished by their structural and physical characteristics. Electrocatalytic hydrogen generation is aided by metal-rich chalcogenide, which Siegmund et al. reported comprises a significant components. $Nb_{21}S_8$ is a sulfide containing many metals [26].

2.3 Classification of chalcogenides according to the number of chalcogen ions

This categorization is based on the relative abundance of chalcogen ions.

2.3.1 Mono-chalcogenides

In the MX structure of mono-chalcogenides, M represents a transition metal and X represents a chalcogenide anion like S_{2-}, Se_{2-}, or Te_{2-}. The surfaces of these compounds are highly durable and have a low coefficient of friction. This is due to their lamellar structure that consists of several thin, stiff sheets formed by strongly bound atoms. Gallium sulfide (GaS) or gallium selenium (GaSe) crystals have been solved, and their designs show two layers of Ga atoms sandwiched between S and Se ions [25].

2.3.2 Dichalcogenides

For instance, MoS_2, $TiSe_2$, MoS_2, and WSe_2 are all examples of metal dichalcogenides, which have the formula MX2, where M = transition metal and X = a chalcogen anion. They crystallize in two-dimensional (2D) structures with a stoichiometry of exactly 1:2. The 2D design consists of X, M, and X sheets held apart by van der Waals forces. The sheets have metal ions that are sixfold coordinated, making a pyramidal shape that is either octahedral or body-centered [16, 25].

2.3.3 Trichalcogenides

Ti, V, Cr, and Mn are only a few early metals found in trichalcogenide groups. M4+ (C2 2-) (C2 -) is a standard notation for these substances, where C = S or Te. One well-known example is niobium triselenide ($NbSe_3$). The acid treatment of tetra-thiomolybdate results in the formation of niobium triselenide [25].

3. Wastewater treatment techniques

Due to the high degree of population growth, the environment quickly becomes polluted and unusable, resulting in severe drought and limited drinking water availability. Municipal and industrial effluents and biological activities represent a severe threat to the quality of water supply, rendering it unfit for human use. For instance, phenols in wastewater are potentially dangerous as they can build up in food chains and become poisonous over time, wreaking havoc on people's health. Heavy metals, suspended particles, pathogens, organic materials, and others are some of the other wastewater components. Conventional procedures, such as physical particle separation, chemical oxygen demand assessment, and biological oxygen demand calculation, which can be used to remove these contaminants and render the water fit for human consumption or other purposes. Large quantities of phosphorus, nitrogen, and other ionic compounds are discharged into the environment and freshwater sources, reducing dissolved oxygen levels in the treated water. Because of the heavy chemical load used to eradicate harmful species before dumping, the water will be carcinogenic, and the local flora and fauna will be decimated. Researchers are developing scalable approaches for fast wastewater treatment, considering current technologies in terms of expense and investment in capital. With its ability to synthesize materials quickly and efficiently and its high effectiveness in eliminating contaminants, nanotechnology is seen as a potential boon to the wastewater treatment and remediation process [27]. Nanoparticles with a diameter of less than 100 nm have a larger surface area than their volume, giving them many active sites and boosting their efficiency. Nanotubes, quantum dots, nanowires, nanocolloids, and nanofilms are only a few of the nanomaterials reported in research. These nanomaterials are crucial for developing nanoparticles with specific properties [28]. Other nanomaterials with physical, chemical, and electrical properties useful for water purification have also been developed. Another option is soft nanomaterials, such as the surfactants, typically employed to regulate coagulation and flocculation [29]. Nanotechnology has allowed us to see the several stages of treatment more clearly, including those involving catalytic oxidation, adsorption and separation, and disinfection [19]. Various treatment methods for wastewater are shown in **Figure 2**.

3.1 Photocatalysis

Groundwater, municipal sewage treatment plant, and surface water have all been contaminated due to the widespread use of PTEs and the problems involved with

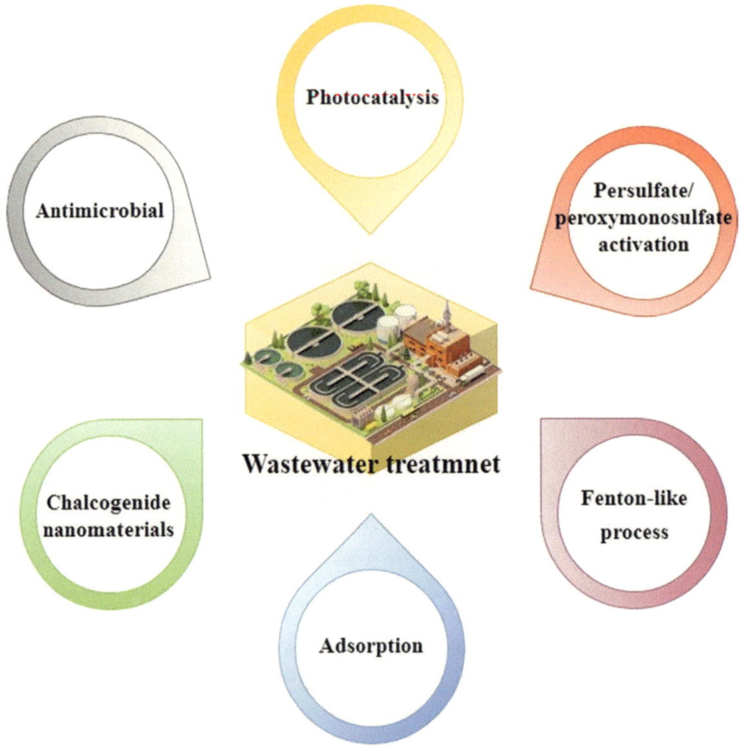

Figure 2.
Wastewater treatment methods.

their eradication from wastewater. PTEs with poor biodegradability are particularly common in wastewater from various industries, including the textile, agricultural, and pharmaceutical sectors [30, 31]. Conventional pollutant removal techniques have been hampered by the difficulty of decomposing and mineralizing these complexes. Solar energy to destroy organic pollutants is known as photocatalysis, an eco-friendly process. Photocatalysis is one of the advanced oxidation processes (AOPs) that have attracted much attention in recent years, especially the discovery of photoelectrochemical water-splitting reactions using semiconductors. Wastewater treatment plants have used this eco-friendly, energy-saving, sustainable technology to eliminate contaminants of high concentration, high complexity, and low biodegradability [32]. Photocatalysts have been demonstrated to be effective in degrading nonbiodegradable organic contaminants in water, according to findings from the literature [33]. Metal oxides like zinc oxide and titanium dioxide (ZnO and TiO_2) have found extensive use as semiconductors in photocatalysis. Reactive oxygen species (ROS), such as the highly oxidizing superoxide radical anion (O_2) and hydroxyl radicals (OH), are produced when the photo-induced electrons and holes react with water (H_2O), the hydroxyl group (OH), and oxygen (O_2). ROS is the driving force behind the deterioration of polar organic compounds (POCs) in wastewater [3].

3.2 Adsorption

As a surface phenomenon, adsorption involves the accumulation of the adsorbate on the absorbent's surface. Adsorption occurs when a solution containing an adsorbable solute comes into contact with a highly porous surface structure, where part of the solute molecules are deposited due to liquid-solid intermolecular forces of attraction [34]. The atoms that make up a typical bulk material provide each other with the necessary bonding interactions, whether covalent, metallic, or ionic. Adsorbents tend to attract adsorbates as their surfaces are not covered by all their atoms [35–37]. Adsorption can be roughly classified as chemisorption (the bonding of adsorbate due to electrostatic attraction), with the precise bonding dynamics depending on the properties of the involved species [38]. Physisorption (attaching the adsorbates to the surface via weak van der Waals forces) and covalent bonding are other possibilities.

3.3 Antimicrobial

Treatment of medications, cosmetics, and colors that remain in the environment requires the employment of nanomaterials with potent antibacterial capabilities [39]. The method by which nanomaterials exert their antibacterial action is poorly understood. However, limited research suggests that this process typically begins with producing reactive oxygen species (ROS), which interact with proteins in the microbial cell wall, denaturing the wall. By breaching the cell wall, ROS can disrupt the microbe's respiratory chain, killing the cell [3]. Several applications involving transition metal chalcogenide nanoparticles contains genetically modified microbes [19], as shown in **Figure 3**.

3.4 Peroxydisulfate/peroxymonosulfate activation

Homogeneous catalysis has been widely used to degrade organic contaminants, but the difficulty of isolating the catalyst has hindered it. Because of their high reaction selectivity, extended lifespan, and broad pH operating range, sulfate radical-based procedures have received much attention [40]. Peroxydisulfate (PDS) and peroxymonosulfate (PMS) can be efficiently activated with transition metals like Cu+; Mn^{2+}; Co^{2+} through electron transfer, light irradiation, and heat to create sulfate radical [41]. Sulfate radicals can be activated in a general way by the metal ions Fe^0, Fe^{2+}, and Fe^{3+} [42].

3.5 Fenton-like process

One method that shows promise and safe for the environment is the heterogeneous Fenton oxidation process used in the waste water treatment. It is a highly desirable

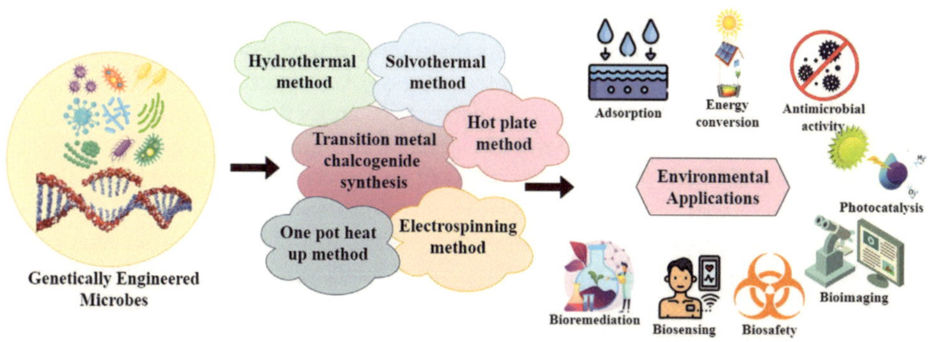

Figure 3.
Genetically engineered microbes in transition metal chalcogenide synthesis for various environmental applications.

technique because of its efficiency, low cost, and simple operating conditions [43, 44]. Heterogeneous Fenton/Fenton-like processes generate hydrogen peroxide (H_2O_2) locally by activating molecular oxygen, unlike the more cumbersome and potentially dangerous H_2O_2 storage required by the traditional Fenton process. Chemical, electrochemical, or photocatalytic processes could all be used to activate oxygen. Catalytic decomposition of locally produced H_2O_2 can result in the formation of $^{\bullet}OH$ using a wide variety of iron-based and iron-free materials [45–47]. Wastewater treatment using the Fenton-like process has used several catalysts based on transition metal oxides (TMOs) and TMCs [46].

4. Chalcogenide nanomaterials

Chalcogen comes from the Greek and Latin words for copper (khalkos) and "born" (genes), respectively. The chalcogenide compounds comprise group VI elements like arsenic, and the elements play a vital role in biosynthetic processes. These chalcogenides are part of the oxygen family of metals, including sulfides, selenides, and tellurides. A chalcogenide chemical compound, amorphous or crystalline, that contains a minimum of one chalcogen anion, such as Se, S, or Te, and a minimum one other electropositive element linked by covalent bonds. Metal oxides, sulfides, selenides, and tellurides based on group VI elements (Zn, Cd, Pb, Cu, Ag, etc.) contain semiconducting chalcogenide binary compounds. However, metal chalcogenides would corrode when exposed to radiation without any electron donors. It employs appropriate band engineering to prevent the rapid recombination of h+/e couples and retrograde reactions. Particle surface modification is essential for enhancing photocatalytic activity by blocking electron-hole recombination. Metal chalcogenide materials can be made better in a few ways. Dopants might establish traps between the photocatalyst's valence band (VB) and conduction band (CB), which could absorb light and prevent recombination by trapping electrons or protons. Dopants can potentially improve charge separation, which is crucial for the photocatalytic reaction [48].

(a) Metal ion: due to their excellent electrical and optical properties, these nanomaterials have many optoelectronic applications in photovoltaics, display devices, thermoelectric generators and refrigerators, lithium-ion batteries, fuel cells, and supercapacitors. (b) Ions of a nonmetal reduce the band gap energy because their impurity states are located near the VB and push the VB edge upward. They exhibit the ability to couple with a wide range of metals and chalcogenides [49]. The polymer sensitizes the surface. The recombination of h+/e− couples can be prevented by using surface imperfections as adsorption sites, where a charge transfer to the adsorbed species occurs. Dye sensitization broadens the photocatalyst's range of excitation energies into the visible area, allowing it to harness solar radiation more effectively. Combining two semiconductors with the correct CB and VB of the photocatalyst can improve the collection of photogenerated electrons and holes on the various semiconductor surfaces and boost the redox reactions of the electrons and holes [50].

Metal chalcogenides are used in various applications, including solar cells, optoelectronic sensors, particle detection and tracking by fluorescent labeling, and even cancer diagnosis [51]. Superconductors, fuel cells, photovoltaics, photocatalysts, and energy storage are exciting uses for chalcogenide nanostructures [52]. Chalcogenide materials are amenable to a wide range of synthesis techniques, including hydrothermal/solvothermal, microwave-assisted, sonochemical, electrochemical, and vacuum synthesis. In addition, chalcogenide materials can be combined with graphene and other forms of advanced carbon [22]. NiS, a sulfide-based chalcogenide, has garnered much attention due to its potential application as a rechargeable lithium battery and its abundance of phases [53]. The utilization of chalcogenide-based materials and other nanomaterials are employed for the removal of various pollutants from the environment, including dyes, chlorinated organic compounds, organophosphorus chemicals, volatile organic compounds, and halogenated herbicides. Environmental pollution and energy constraint are two of the major environmental problems caused by the expansion of industrial and technical sectors. For this reason, it is critical to combat the negative effects on the environment by creating novel nanocomposites that can mediate such effects. Catalysts based on transition metal chalcogenides have being studied for potential uses in various fields because of their stability, optoelectronic behavior, and indirect band gaps, which allow them to absorb visible light, which is abundant in solar radiation [54].

Metal chalcogenide nanostructures are being considered for use in various sensors, light-emitting diodes, Li-ion batteries, solar cells, supercapacitors, thermoelectric devices, fuel cells, and storage devices. In recent years, metal chalcogenides have become popular as potential components in solar absorber devices. The well-known quantum size effect causes metal chalcogenides' physical and chemical properties to emerge at the nanoscale, where they had previously been hidden from view. In addition, the specific surface area of nanostructured metal chalcogenides is much higher than that of their bulk counterparts. It is helpful for energy devices since it facilitates a more rapid reaction or interaction between them and the interacting medium. Nanostructured metal chalcogenides, in contrast to their bulk counterparts, can have a significantly higher specific surface area, which is beneficial for energy devices since it allows for a faster reaction/interaction among the components and their surrounding medium [50]. Developing new nanostructured materials of varying sizes and morphologies has benefited the many uses of metal chalcogenides. Nanostructured materials have several benefits over their bulk equivalents when used as a catalyst or electrode material in electrochemical storage systems, including (1) Higher specific surface areas and more active sites, (2) increased durability after repeated uses, (3) enhanced electrical conductivity, and (4) decreased electron transport route lengths.

4.1 Chalcogenides that are metal-based

Cadmium sulfide has gained much interest because of its remarkable chemical, physical, and optical features; nevertheless, it has a high recombination rate. As a result, Aboud et al.'s spray pyrolysis technique for synthesizing Cu-doped CdS to lower the recombination rate was successfully implemented [55].

4.2 Chalcogenides containing transition metals

Cocatalysts made of noble metals like Pt have been found to boost the photocatalytic efficiency of chalcogenide materials, particularly CdS [56]. Rapid one-pot solvothermal synthesis technique yields excellent Pt/CdS photocatalyst, as described by Fu et al. The photocatalytic activity was reported to increase when exposed to visible light significantly [57]. Solvothermal processing was used to make the materials [56, 57].

4.3 Chalcogenide composites

MoS_2/CdS is one of the many chalcogenide composites that have been produced and found to be effective as photocatalysts. A simple solvothermal approach was used in the MoS_2/CdS nanocomposite synthesis by He et al. It has been observed that the photocatalytic activity of the synthesized MoS_2/CdS is five times that of pure CdS. The nanocomposite of CdS and graphene is yet another instance. According to Khan et al., the substance was synthesized utilizing a simple and one-step precipitation technique. CdS-graphene nanocomposites increased photocatalytic activity under visible-light irradiation [58].

5. Synthesis of chalcogenide nanomaterials

Several techniques, such as bottom-up synthesis by vapor-phase deposition, solvothermal method, and top-down strategy via micromechanical cleavage, have been developed to manufacture 2D metal chalcogenides. The advent of nanotechnology opened up a wide variety of new doors for several fields, including public health, medicine, environmental protection, and early detection and diagnosis. The commercialization of chalcogenide nanostructures exposes society to a range of risks that are difficult to predict without extensive research. Using chalcogenide, nanomaterials have benefits and drawbacks, just like any other nanomaterial. **Table 2** highlights the advantages and disadvantages of chalcogenide nanoparticles selectively. Therefore, it is essential to regularly choose suitable microorganisms and monitor and adjust variables, such as pH, temperature/incubation temperature, metal ion concentration, and biomass content. During the development phase, microbes oxidize organic compounds using chalcogen oxyanions, which are reduced to sulfur oxyanions, selenium oxyanions, and tellurium oxyanions, respectively [51].

6. Application in wastewater treatment

Environmental, medicinal, materials science, and technological fields have all shown interest in chalcogenide nanoparticles over the past two decades. The chalcogenides are widely regarded as a great material class with numerous promising applications in science and technology. Chalcogenide particles exhibit tunable quantum efficiencies, long photostability, narrow emission spectra, and continuous absorption. Nanocomposites of metal chalcogenides and graphene, as well as

Benefits	Drawbacks
Low-temperature, high-pressure synthesis	Inadequate dimension formation
Easy alteration	The shape complicates the process of achieving the intended shape
Reduced use of energy	Size and shape control is challenging and frequently unreliable
Uniformity	Microbe generation is slower than chemical synthesis; thus, it takes more time
Electron donors are made from sustainable resources.	Expensive
Highly transparent crystals and a high rate of production	A laborious process is being followed

Table 2.
Nanomaterials that have been synthesized: their benefits and drawbacks.

core-shell materials composed of metal chalcogenides, have demonstrated a wide range of uses beyond composites. Hydrogen purification heterojunctions involving various chalcogenide semiconductor materials (e.g., CdS, CdSe, Bi2S3, Cu2S, Ag2S, and CuI) and noble metals (Au and Ag) have been the subject of the most research. For different purposes, scientists looked into chalcogenide-based nanoparticles made by reducing selenite using *Veillonella atypica*. Selenium nanospheres can be produced by the anaerobic bacterium *Veillonella atypica* by reducing selenium oxyanions. Bacteria can further decrease selenium nanospheres to reactive selenide. This reactive selenide can be precipitated with a metal cation to create nanoscale chalcogenide precipitates, like zinc selenide, which potentially have optical and semiconducting properties. When reducing selenite, the entire cell population relied on hydrogen as an electron donor; adding a redox mediator, such as anthraquinone disulfonic acid, accelerated the reduction rate [19].

Degradation of pollutants, photoreduction of pigments and dyes, and photocatalysts are just some of the many current applications of photocatalysis. Other applications include photocatalysis, solar cells, photochemical reactions applications, optoelectronic materials for LEDs, etc. [19]. The uses of chalcogenide nanoparticles range widely and include the following: Biochemical infrared sensing with selenide glass fibers, ZnSe as an infrared (IR) window IR sensor, solar panels comprised solar cells made of polymer, and sensors that use electrochemistry. Typically, the practical

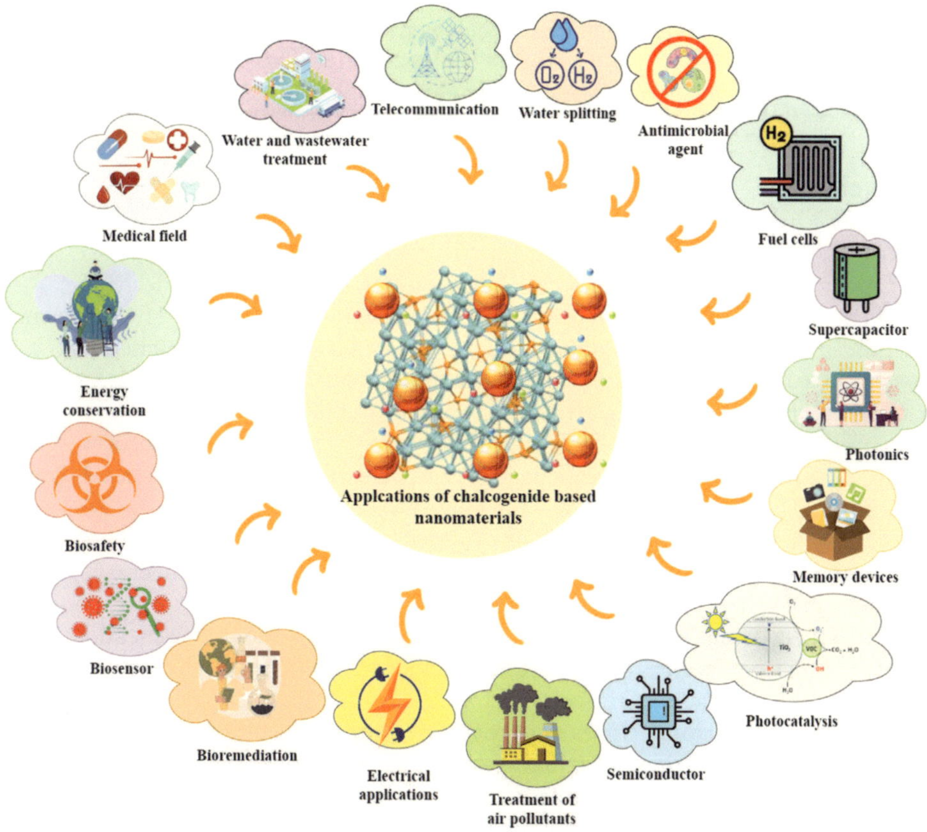

Figure 4.
Applications of chalcogenide-based nanomaterials in various fields.

applicability of binary metal chalcogenides is limited due to their reduced charge separation efficiency and susceptibility to photocorrosion. Chalcogenide-based nanomaterials find applications in a wide variety of activities, such, uses in medicine, imaging using X-rays, decontamination by microorganisms, metabolic profiling: a tool for studying organisms of all kinds, disposal of heavy metals, electrocatalysis, treatment of wastewater, managing infectious disease-causing organisms, and organic matter breakdown. The applications of chalcogenide-based nanomaterials in various fields are shown in **Figure 4**.

7. Challenges and prospects of chalcogenides

For the future sustainable and precise manufacture and usage of chalcogenide-based nanomaterials for practical applications, the following difficulties need to be addressed from a photographic perspective:

- For chalcogenide-based nanomaterials, size, shape, and monodispersity are crucial criteria. To achieve the desired optical properties of chalcogenide composites, it is necessary to systematically investigate and confirm methods for effective control of the particle size distribution, morphology, and monodispersity by varying the synthetic parameters and solvents used.

- By designing heterojunction metal chalcogenides that are multirich electrons and the separation of charge carriers (holes and electrons), one can achieve the desired optical properties of chalcogenide composites.

8. Conclusion

Sulfide, selenide, and telluride-based chalcogenides are typical examples of chalcogenides. Many technologies use chalcogenide materials or their derivatives, such as those based on binary, ternary, and quaternary chalcogenides. Hydrothermal, one-pot, solvothermal, sonochemical, microwave-assisted, and electrochemical processes are few of the ways chalcogenide compounds can be synthesized. Raw material composition, synthesis strategies, and post-synthesis processing all have influenced the morphology of chalcogenide materials. An intriguing platform for light-harvesting applications are using chalcogenides and chalcogenide-based nanomaterials with smart hybridization. These novel materials show great promise for photocatalytic water-splitting reactions to generate hydrogen gas (H_2), degrading organic and inorganic contaminants, reducing carbon dioxide (CO_2) to create renewable fuels, and other solar-to-energy conversion processes.

Due to their potential usefulness in various technological contexts, metal chalcogenides have recently gained attention as a significant material class. The immobilization of nanomaterials on reactor surfaces necessitates paying close attention to trapping the nanoparticles onto specific membrane surfaces to prevent them from escaping and entering the drinking water due to the risks they bring to human health and the environment. To isolate and keep the nanoparticles in suspension, new research approaches should be developed to bind them onto the reactor/membrane surfaces firmly. During the environmental rehabilitation process, these chalcogenide-based compounds were used as catalysts for removing various contaminants and converting industrial effluents into drinkable water. Remediation-related chalcogenide-based nanoparticles are produced by *Veillonella atypica* through selenite reduction. Severe environmental problems, such as pollution and a lack of renewable energy, have resulted directly from progress in the industrial and other sectors. To counteract these unfavorable environmental effects, novel nanocomposites must be developed and used in ecological restoration. Transition metal chalcogenides were utilized even in highly specialized environmental remediation applications, such as dealing with radionuclides found in legacy wastes from nuclear weapons development or wastes generated during nuclear power generation or nuclear fuel reprocessing.

The following suggestions, are essential for promoting the use of TMOs, TMCs, and their composites in water and wastewater treatment operations. It is not scientifically sound to utilize any poisonous material to remove another harmful material; hence, more research is needed to analyze the toxicity of the TMOs and TMCs and their composites. For commercial uses of TMOs and TMCs, it is necessary to produce them on a large scale, but this cannot be done without conducting a scientific investigation to determine the best synthesis methods. Surface modifications to TMOs and TMCs can lessen charge carrier recombination, but additional research into the recombination mechanism is necessary before photocatalysis can be optimized for broader uses.

Author details

Arunkumar Priya[1,2] and Suresh Sagadevan[3*]

1 Department of Chemical Engineering, KPR Institute of Engineering and Technology, Tamilnadu, India

2 Centre for Nanoscience and Technology, KPR Institute of Engineering and Technology, Tamilnadu, India

3 Nanotechnology and Catalysis Research Centre, Universiti Malaya, Kuala Lumpur, Malaysia

*Address all correspondence to: drsureshsagadevan@um.edu.my

References

[1] Ahmad T, Aadila RM, Ahmeda H, Rahman U, Soares BCV, Souza SLQ, et al. Treatment and utilization of dairy industrial waste: A review. Trends in Food Science and Technology. 2019;**88**:361-372

[2] Madakkaruppan V, Pius A, Sreenivas T, Sunilkumar TS. Behaviour of Si, Al, Fe and Mg during oxidative sulfuric acid leaching of low grade uranium ore: A kinetic approach. Journal of Environmental Chemical Engineering. 2019;**7**:103139

[3] Okpara EC, Olatunde OC, Wojuola OB, Onwudiwe DC. Applications of transition metal oxides and chalcogenides and their composites in water treatment: A review. Environmental Advances. 2023;**11**:100341. DOI: 10.1016/j.envadv.2023.100341

[4] Okpara EC, Ajiboye TO, Onwudiwe DC, Wojuola OB. Optical and electrochemical techniques for point-of-care water quality monitoring: A review. Results in Chemistry. 2022:100710

[5] Che NS, Bett S, Okpara EC, Olagbaju PO, Fayemi OE, Mathuthu M. An assessment of land use and land cover changes and its impact on the surface water quality of the Crocodile River catchment, South Africa. In: River Deltas-Recent Advances. London, UK: IntechOpen; 2021

[6] Okpara EC, Fayemi OE, Sherif E-SM, Ganesh PS, Swamy BEK, Ebenso EE. Electrochemical evaluation of Cd2+ and Hg2+ ions in water using ZnO/Cu2ONPs/PANI modified SPCE electrode. Sensing and Bio-Sensing Research. 2022;**35**:100476

[7] Sadegh H, Ali GAM, Gupta VK, Makhlouf ASH, Shahryari-Ghoshekandi R, Nadagouda MN, et al. The role of nanomaterials as effective adsorbents and their applications in wastewater treatment. Journal of Nanostructure in Chemistry. 2017;**7**:1-14

[8] Aigbe UO, Ukhurebor KE, Onyancha RB, Osibote OA, Darmokoesoemo H, Kusuma HS. Fly ash-based adsorbent for adsorption of heavy metals and dyes from aqueous solution: A review. Journal of Materials Research and Technology. 2021;**14**:2751-2774

[9] Dutta V, Singh P, Shandilya P, Sharma S, Raizada P, Saini AK, et al. Review on advances in photocatalytic water disinfection utilizing graphene and graphene derivatives-based nanocomposites. Journal of Environmental Chemical Engineering. 2019;**7**:103132

[10] Macaskie LE, Mikheenko IP, Yong P, Deplanche K, Murray AJ, Paterson-Beedle M, et al. Today's wastes, tomorrow's materials for environmental protection. Advances in Materials Research. 2009;**71**:541-548

[11] Jeevanandam J, Barhoum A, Chan YS, Dufresne A, Danquah MK. Review on nanoparticles and nanostructured materials: History, sources, toxicity and regulations. Beilstein Journal of Nanotechnology. 2018;**9**:1050-1074

[12] Barrak H, Saied T, Chevallier P, Laroche G, M'nif A, Hamzaoui AH. Synthesis, characterization, and functionalization of ZnO nanoparticles by N-(trimethoxysilylpropyl) ethylenediamine triacetic acid (TMSEDTA): Investigation of the interactions between Phloroglucinol and ZnO@ TMSEDTA. Arabian Journal of Chemistry. 2019;**12**:4340-4347

[13] Gao M-R, Xu Y-F, Jiang J, Yu S-H. Nanostructured metal chalcogenides: Synthesis, modification, and applications in energy conversion and storage devices. Chemical Society Reviews. 2013;**42**:2986-3017

[14] Gao M, Jiang J, Yu S. Solution-based synthesis and design of late transition metal chalcogenide materials for oxygen reduction reaction (ORR). Small. 2012;**8**:13-27

[15] Zhou M, Xiao K, Jiang X, Huang H, Lin Z, Yao J, et al. Visible-light-responsive chalcogenide photocatalyst Ba2ZnSe3: Crystal and electronic structure, thermal, optical, and photocatalytic activity. Inorganic Chemistry. 2016;**55**:12783-12790

[16] Khan MM. Introduction and fundamentals of chalcogenides and chalcogenides-based nanomaterials. In: Chalcogenide-Based Nanomaterials as Photocatalysis. Elsevier; 2021. pp. 1-6

[17] Zhou X, Rodriguez EE. Tetrahedral transition metal chalcogenides as functional inorganic materials. Chemistry of Materials. 2017;**29**:5737-5752

[18] Xiao J-R, Yang S-H, Feng F, Xue H-G, Guo S-P. A review of the structural chemistry and physical properties of metal chalcogenide halides. Coordination Chemistry Reviews. 2017;**347**:23-47

[19] Enamala MK, Tangellapally A, Chavali M, Pasumarthy D, Murthy MK, Kuppam C, et al. Use of Chalcogenides-Based Nanomaterials for Wastewater Treatment Including Bacterial Disinfection and Organic Contaminants Degradation. Elsevier Inc.; 2021. DOI: 10.1016/B978-0-12-820498-6.00010-X

[20] Khan MM. Introduction and Fundamentals of Chalcogenides and Chalcogenides-Based Nanomaterials. Elsevier Inc.; 2021. DOI: 10.1016/B978-0-12-820498-6.00001-9

[21] Bajpai PK, Yadav S, Tiwari A, Virk HS. Recent advances in the synthesis and characterization of chalcogenide nanoparticles. Solid State Phenomena. 2015;**222**:187-233

[22] Queffurus M, Barnes S-J. A review of sulfur to selenium ratios in magmatic nickel–copper and platinum-group element deposits. Ore Geology Reviews. 2015;**69**:301-324

[23] Schorr S. Structural aspects of adamantine like multinary chalcogenides. Thin Solid Films. 2007;**515**:5985-5991

[24] Medina-Ramírez I, Hernández-Ramírez A, Maya-Trevino ML. Synthesis methods for photocatalytic materials. In: Photocatalytic Semiconductors Synthesis Characterization, and Environmetal Applications. 2015. pp. 69-102

[25] Ahluwalia GK. Applications of Chalcogenides: S, Se, and Te. 2017

[26] Siegmund D, Blanc N, Smialkowski M, Tschulik K, Apfel U. Metal-rich chalcogenides for electrocatalytic hydrogen evolution: Activity of electrodes and bulk materials. ChemElectroChem. 2020;**7**:1514-1527

[27] Nnaji CO, Jeevanandam J, Chan YS, Danquah MK, Pan S, Barhoum A. Engineered nanomaterials for wastewater treatment: Current and future trends. In: Fundamental Nanoparticles. 2018. pp. 129-168

[28] Fadlalla MI, Senthil Kumar P, Selvam V, Ganesh BS. Recent advances in nanomaterials for wastewater treatment. In: Advanced Nanostructured Materials

for Environmetal Remediation. 2019. pp. 21-58

[29] Zhao W, Chen I-W, Huang F. Toward large-scale water treatment using nanomaterials. Nano Today. 2019;**27**:11-27

[30] Kushniarou A, Garrido I, Fenoll J, Vela N, Flores P, Navarro G, et al. Solar photocatalytic reclamation of agro-waste water polluted with twelve pesticides for agricultural reuse. Chemosphere. 2019;**214**:839-845

[31] Olama N, Dehghani M, Malakootian M. The removal of amoxicillin from aquatic solutions using the TiO 2/UV-C nanophotocatalytic method doped with trivalent iron. Applied Water Science. 2018;**8**:1-12

[32] Guo Y, Qi PS, Liu YZ. A review on advanced treatment of pharmaceutical wastewater. In: IOP Conference Series: Earth and Environmental Science. Vol. 63. IOP Publishing; 2017. p. 12025

[33] Hak CH, Sim LC, Leong KH, Lim PF, Chin YH, Saravanan P. M/gC 3 N 4 (M= Ag, Au, and Pd) composite: Synthesis via sunlight photodeposition and application towards the degradation of bisphenol A. Environmental Science and Pollution Research. 2018;**25**:25401-25412

[34] Huang J-J, Lin C-C, Wuu D-S. Antireflection and passivation property of titanium oxide thin film on silicon nanowire by liquid phase deposition. Surface and Coatings Technology. 2017;**320**:252-258

[35] Micheal K, Ayeshamariam A, Boddula R, Arunachalam P, AlSalhi MS, Theerthagiri J, et al. Assembled composite of hematite iron oxide on sponge-like BiOCl with enhanced photocatalytic activity. Materials Science for Energy Technologies. 2019;**2**:104-111

[36] Jo YK, Lee JM, Son S, Hwang S-J. 2D inorganic nanosheet-based hybrid photocatalysts: Design, applications, and perspectives. Journal of Photochemistry and Photobiology. 2019;**40**:150-190

[37] Chiu Y-H, Chang T-FM, Chen C-Y, Sone M, Hsu Y-J. Mechanistic insights into photodegradation of organic dyes using heterostructure photocatalysts. Catalysts. 2019;**9**:430

[38] Cheng C, Liang Q, Yan M, Liu Z, He Q, Wu T, et al. Advances in preparation, mechanism and applications of graphene quantum dots/semiconductor composite photocatalysts: A review. Journal of Hazardous Materials. 2022;**424**:127721

[39] Siddiqui SI, Manzoor O, Mohsin M, Chaudhry SA. Nigella sativa seed based nanocomposite-MnO2/BC: An antibacterial material for photocatalytic degradation, and adsorptive removal of methylene blue from water. Environmental Research. 2019;**171**:328-340

[40] Zhu K, Jin C, Zhao C, Hu R, Klencsar Z, Sundaram GA, et al. Modulation synthesis of multi-shelled cobalt-iron oxides as efficient catalysts for peroxymonosulfate-mediated organics degradation. Chemical Engineering Journal. 2019;**359**:1537-1549

[41] Hassani A, Eghbali P, Kakavandi B, Lin K-YA, Ghanbari F. Acetaminophen removal from aqueous solutions through peroxymonosulfate activation by CoFe2O4/mpg-C3N4 nanocomposite: Insight into the performance and degradation kinetics. Environmental Technology and Innovation. 2020;**20**:101127

[42] Monga D, Shetti NP, Basu S, Reddy KR, Badawi M, Bonilla-Petriciolet A, et al. Engineered biochar:

A way forward to environmental remediation. Fuel. 2022;**311**:122510

[43] Yang Z, Zhang X, Pu S, Ni R, Lin Y, Liu Y. Novel Fenton-like system (Mg/Fe-O2) for degradation of 4-chlorophenol. Environmental Pollution. 2019;**250**:906-913

[44] Zhou W, Meng X, Gao J, Alshawabkeh AN. Hydrogen peroxide generation from O2 electroreduction for environmental remediation: A state-of-the-art review. Chemosphere. 2019;**225**:588-607

[45] Su P, Zhou M, Lu X, Yang W, Ren G, Cai J. Electrochemical catalytic mechanism of N-doped graphene for enhanced H2O2 yield and in-situ degradation of organic pollutant. Applied Catalysis B: Environmental. 2019;**245**:583-595

[46] Liu Y, Zhao Y, Wang J. Fenton/Fenton-like processes with in-situ production of hydrogen peroxide/hydroxyl radical for degradation of emerging contaminants: Advances and prospects. Journal of Hazardous Materials. 2021;**404**:124191

[47] Gao S, Feng D, Chen F, Shi H, Chen Z. Multi-functional well-dispersed pomegranate-like nanospheres organized by ultrafine ZnFe2O4 nanocrystals for high-efficiency visible-light-Fenton catalytic activities. Colloids and Surfaces A: Physicochemical and Engineering Aspects. 2022;**648**:129282

[48] Stronski A, Paiuk O, Gudymenko A, Klad'Ko V, Oleksenko P, Vuichyk N, et al. Effect of doping by transitional elements on properties of chalcogenide glasses. Ceramics International. 2015;**41**:7543-7548

[49] Theerthagiri J, Karuppasamy K, Durai G, Rana AUHS, Arunachalam P, Sangeetha K, et al. Recent advances in metal chalcogenides (MX; X= S, Se) nanostructures for electrochemical supercapacitor applications: A brief review. Nanomaterials. 2018;**8**:256

[50] Sarma GVSS, Chavali M, Nikolova MP, Enamala MK, Kuppan C. Basic Principles, Fundamentals, and Mechanisms of Chalcogenide-Based Nanomaterials in Photocatalytic Reactions. Elsevier Inc.; 2021. DOI: 10.1016/B978-0-12-820498-6.00004-4

[51] Mal J, Nancharaiah YV, Van Hullebusch ED, Lens PNL. Metal chalcogenide quantum dots: Biotechnological synthesis and applications. RSC Advances. 2016;**6**:41477-41495

[52] Muslih EY, Munir B, Khan MM. Advances in Chalcogenides and Chalcogenides-Based Nanomaterials Such as Sulfides, Selenides, and Tellurides. Elsevier Inc.; 2021. DOI: 10.1016/B978-0-12-820498-6.00002-0

[53] Han S-C, Kim K-W, Ahn H-J, Ahn J-H, Lee J-Y. Charge–discharge mechanism of mechanically alloyed NiS used as a cathode in rechargeable lithium batteries. Journal of Alloys and Compounds. 2003;**361**:247-251. DOI: 10.1016/S0925-8388(03)00380-3

[54] Shanmugaratnam S, Rasalingam S. Transition metal chalcogenide (TMC) nanocomposites for environmental remediation application over extended solar irradiation. In: Nanocatalysts. London, UK: IntechOpen; 2019. p. 75

[55] Aboud AA, Mukherjee A, Revaprasadu N, Mohamed AN. The effect of Cu-doping on CdS thin films deposited by the spray pyrolysis technique. Journal of Materials Research and Technology. 2019;**8**:2021-2030

[56] Choi YI, Lee S, Kim SK, Kim Y-I, Cho DW, Khan MM, et al. Fabrication of ZnO, ZnS, Ag-ZnS, and Au-ZnS microspheres for photocatalytic activities, CO oxidation and 2-hydroxyterephthalic acid synthesis. Journal of Alloys and Compounds. 2016;**675**:46-56

[57] Yuan Y-J, Chen D, Yu Z-T, Zou Z-G. Cadmium sulfide-based nanomaterials for photocatalytic hydrogen production. Journal of Materials Chemistry A. 2018;**6**:11606-11630

[58] He G, Zhang Y, He Q. MoS2/CdS heterostructure for enhanced photoelectrochemical performance under visible light. Catalysts. 2019;**9**:379

Chapter 5

Nano to Macro Production and Applications of Chalcogenides

Manivel Rajan, Raja Arumugam, Sivasubramani Vediyappan, Siva Vadivel, Rajesh Paulraj and Ramasamy Perumalsamy

Abstract

Chalcogenides are basically one chalcogen anion with a more electropositive cation. Selenium, Tellurium and Sulfur based chalcogenides are used widely with a variety of applications. Chalcogenides are known as an IR transmission and high reflective index, with a wide range of applications in catalyst technologies and sensing devices. It is possible to make chalcogenides in various forms like nanocrystals, thin films and bulk crystals based on the requirement. Chalcogenides are categorized as binary (2°), ternary (3°), quaternary (4°), and penternary (5°) based on their structural differences. These compounds have a high degree of versatility for modifying the bandgap without the use of hazardous components. The structural and chemical property analysis will help us to tailor the chalcogenides-based material for the suitable application and reveal the science behind this important class of materials. The diverse size synthesis of chalcogenides, encompassing nano, micro, and macro scales, is crucial for tailoring their properties to meet specific applications, ranging from nanoscale innovations in quantum dots for advanced electronics to microscale developments in thin-film solar cells for efficient photovoltaics, and macroscale applications in solid-state memory devices and radiation detectors, showcasing the versatile impact of size-tailored chalcogenides across a spectrum of technologies.

Keywords: chalcogenides, physical properties, multiscale production, versatile application, electronics

1. Introduction

Chalcogenides, compounds composed of chalcogens (Group 16 elements like sulfur, selenium, and tellurium) with metals or metalloids, possess a range of remarkable physical properties [1]. Their electrical characteristics can vary widely, with some chalcogenides exhibiting metallic conductivity, while others are semiconductors or insulators. Many chalcogenides have a high refractive index, making them valuable in optical devices. They often exhibit non-linear optical behavior, such as second harmonic generation, making them vital in photonic applications [2]. Chalcogenides are also known for their ability to undergo phase transitions, switching between amorphous and crystalline states, which is crucial in data storage technologies. These materials can be engineered to exhibit superconductivity, high thermoelectric efficiency, or even ferroelectric properties. Additionally, their strong bonding capabilities and ability to form stable glasses enable their use in diverse applications, including infrared optics, radiation detection, and semiconductor devices. Chalcogenides' wide-ranging physical properties make them essential in a multitude of scientific and industrial contexts, from electronics and optics to energy conversion and memory technologies. **Figure 1** shows the place of chalcogenides in periodic table.

Chalcogenides exhibit diverse structural properties owing to their unique compositions and bonding characteristics [3]. These materials can form a variety of crystal structures, including simple cubic, hexagonal, and tetragonal, depending on the specific chalcogen and metal elements involved. The presence of lone pair electrons in chalcogens often leads to distortions in their crystal structures and the formation of layered or chain-like arrangements. Moreover, chalcogenides can transition between amorphous and crystalline states, a property crucial for their application in phase-change memory devices [4]. These structural features are central to the wide-ranging utility of chalcogenides in fields like solid-state physics, materials science, and electronics, where their structural adaptability allows for tailoring their properties to meet specific requirements.

The bonding properties of chalcogenides are a defining characteristic of these compounds, stemming from the interactions between chalcogens (Group 16 elements) and metals or metalloids [5]. Chalcogenides typically exhibit covalent or polar covalent bonds, which involve the sharing of electrons between the chalcogen and the metalloid. The nature of these bonds can lead to unique properties like strong optical responses, as chalcogenides often possess high refractive indices and non-linear optical behavior. Chalcogenides' ability to form stable glasses is attributed to the covalent bonding network, making them valuable in optical and infrared applications. Moreover, the presence of lone pair electrons in chalcogens can result in polar bonds, impacting the structural and electronic properties of these materials. These

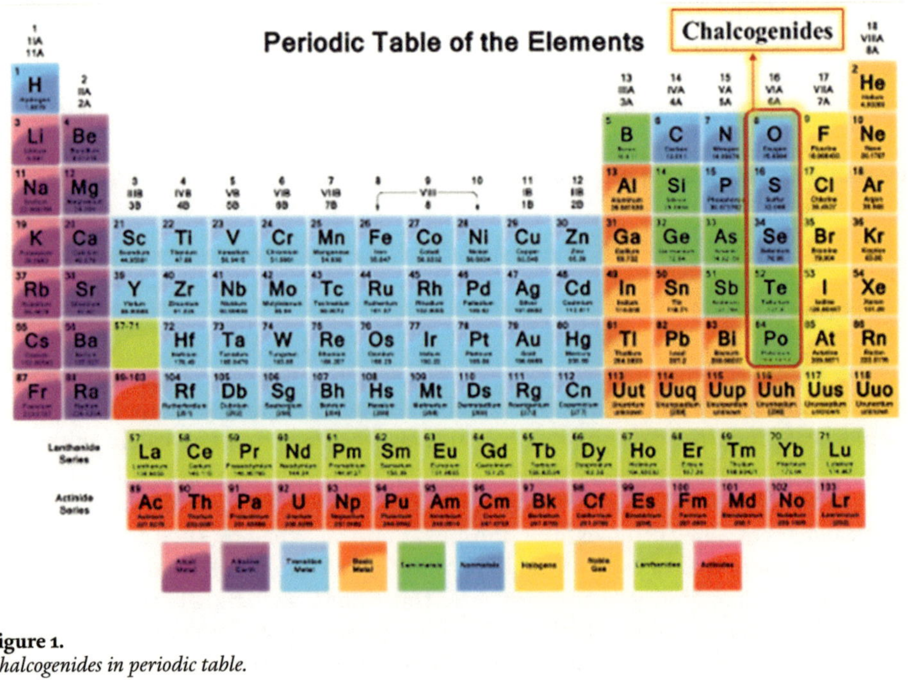

Figure 1.
Chalcogenides in periodic table.

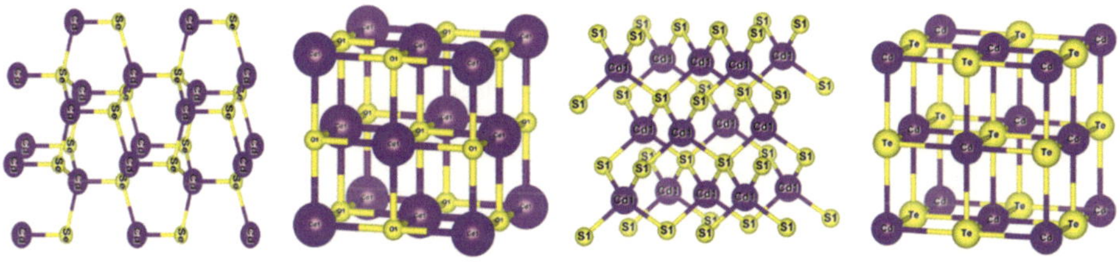

Figure 2.
Different bonding and structure in chalcogenides.

bonding properties are at the heart of chalcogenides' versatility, enabling their use in a wide range of applications, from semiconductor devices to optical components and data storage technologies. **Figure 2** represents the types of bonding and structure of chalcogen elements with cadmium atom.

The optical properties of chalcogenides are of paramount importance due to their wide-ranging applications in optics, photonics, and infrared technology [6]. Chalcogenides typically exhibit a high refractive index, allowing for efficient light confinement and propagation. Their ability to efficiently transmit infrared radiation, often spanning the mid- and far-infrared regions, makes them vital in thermal imaging devices, night vision technology, and sensing applications. Chalcogenide glasses can display non-linear optical behavior, such as second harmonic generation and four-wave mixing, making them invaluable in photonic applications like optical amplifiers and frequency converters. Additionally, chalcogenides are known for their strong infrared absorption characteristics, which are exploited in the creation of optical components like lenses, windows, and fibers for use in telecommunications, spectroscopy, and thermal imaging systems. Their remarkable optical properties underpin the versatility and significance of chalcogenides in various fields, ranging from telecommunications to defense and medical imaging.

The electrical properties of chalcogenides are highly diverse and can vary widely depending on their composition and structure. Many chalcogenides are semiconductors, meaning they have intermediate electrical conductivity, and their electrical behavior can be manipulated through doping or alloying [7]. Some chalcogenides, like bismuth telluride (Bi_2Te_3), exhibit excellent thermoelectric properties, efficiently converting heat differentials into electricity. On the other hand, certain chalcogenides are metallic in nature, offering high electrical conductivity. Notably, phase-change chalcogenides can rapidly switch between amorphous and crystalline states, providing the basis for non-volatile memory devices. The ability to tailor the electrical properties of chalcogenides makes them essential in a wide array of applications, including electronics, thermoelectric power generation, memory technologies, and beyond. Their adaptability and unique electrical characteristics are central to their enduring significance in science and industry.

The transport properties of chalcogenides are a fundamental aspect of their behavior and play a critical role in various applications. These materials exhibit diverse transport characteristics, including electrical conductivity, thermal conductivity, and carrier mobility, which can be tuned by modifying their composition and structure [8]. Chalcogenides like bismuth telluride (Bi_2Te_3) and lead telluride (PbTe) are renowned for their exceptional thermoelectric properties, enabling efficient conversion of heat into electricity, which is valuable for power generation and cooling systems. Additionally, chalcogenides like molybdenum disulfide (MoS_2)

and tungsten di selenide (WSe_2) exhibit intriguing electronic and thermal transport properties, making them essential in the development of two-dimensional materials for electronic and optoelectronic applications. Understanding and manipulating the transport properties of chalcogenides are central to harnessing their full potential in areas such as energy conversion, semiconductors, and emerging nanotechnologies.

2. Various forms of chalcogenides and its production

Chalcogenides are synthesized across a wide spectrum of production sizes, from the nanoscale to the macroscale, offering a versatile range of materials for diverse applications. At the nanoscale, chalcogenides are engineered as nanoparticles, quantum dots, and nanowires, finding utility in nanoelectronics, quantum computing, and advanced sensors. In the microscale realm, chalcogenides are employed in the fabrication of integrated circuits, optical devices, and infrared optics, enabling high-performance telecommunications and thermal imaging. On a macroscopic level, these compounds are utilized in the production of thin-film solar cells, semiconductor devices, and superconductors, contributing to renewable energy, data storage, and cutting-edge technologies, demonstrating their adaptability and significance across various industries and scales.

2.1 Nano scale synthesis

Nanoscale synthesis of chalcogenides involves precision techniques tailored to create materials at the atomic and molecular levels. Common methods include chemical precipitation, where chalcogen precursors react with metal salts in a solvent to produce chalcogenide nanoparticles through controlled chemical reactions. The sol-gel method employs hydrolysis and condensation of precursor compounds in a solution, yielding highly uniform nanomaterials and thin films. Hydrothermal synthesis occurs in high-temperature, high-pressure aqueous environments, allowing for well-defined chalcogenide nanostructures with precise morphologies. Chemical vapour deposition (CVD) is instrumental for depositing chalcogenide nanoparticles and thin films on substrates by thermally decomposing precursor gases, granting researchers exact control over nanostructure growth on surfaces. These nanoscale synthesis techniques are fundamental to applications in nanoelectronics, quantum dots, sensors, and other cutting-edge technologies, where tailored nanomaterial properties are essential.

2.1.1 Chemical precipitation

Chemical precipitation is a widely used method in chemistry for the creation of solid particles from dissolved substances by introducing a chemical reaction that leads to the formation of insoluble products [9]. The process is shown in **Figure 3**. A classic example is the synthesis of metal hydroxides by adding a base, such as sodium hydroxide (NaOH), to a metal salt solution, resulting in the precipitation of metal hydroxides. For instance, when sodium hydroxide is added to a solution containing copper sulphate ($CuSO_4$), it forms copper hydroxide, which precipitates as a solid. This technique is essential in various applications, including water treatment to remove impurities, the synthesis of nanoparticles, and the recovery of valuable metal ions from industrial wastewater. Chemical precipitation allows for efficient separation and purification processes, making it a valuable tool in both laboratory settings and industrial applications.

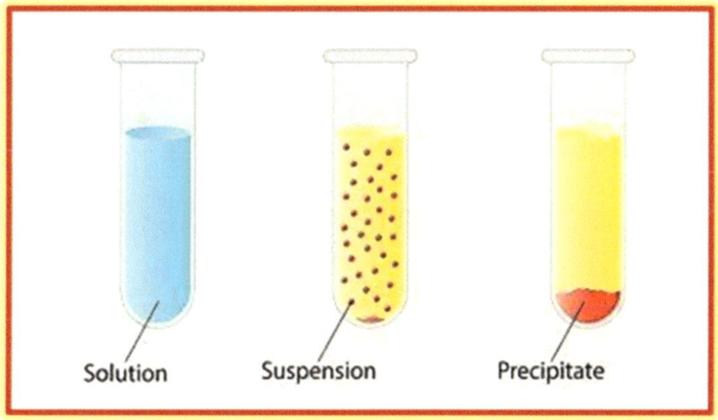

Figure 3.
Precipitation method.

2.1.2 Sol-gel method

The sol-gel method stands as a versatile technique in chalcogenide materials science, particularly in the production of chalcogenide glasses and thin films. **Figure 4** shows the process of sol-gel method. This method involves the synthesis of chalcogenide compounds, combining elements like sulfur, selenium, or tellurium with metal or metalloid components, in a precursor solution. Through a precisely controlled chemical process, this solution transforms into a sol, forming a colloidal suspension of nanoscale chalcogenide particles [10]. This sol is subsequently gelled, typically by initiating further chemical reactions, resulting in the creation of a three-dimensional network of chalcogenide material. An exemplary application of the sol-gel method is the production of chalcogenide glass fibres. By using this approach, the material's composition, structure, and optical properties can be tailored with precision, leading to the production of high-quality chalcogenide optical fibers essential for infrared optics, telecommunications, and sensing applications. The sol-gel technique enables meticulous control over chalcogenide properties, making it a fundamental method in the development of advanced photonic devices and optical components.

2.1.3 Hydrothermal synthesis

Hydrothermal synthesis is a method employed in the production of chalcogenides, particularly chalcogenide nanomaterials and crystals, by subjecting precursor solutions to high-temperature and high-pressure conditions in an autoclave as shown in **Figure 5**. This technique is notably used in the synthesis of cadmium selenide (CdSe)

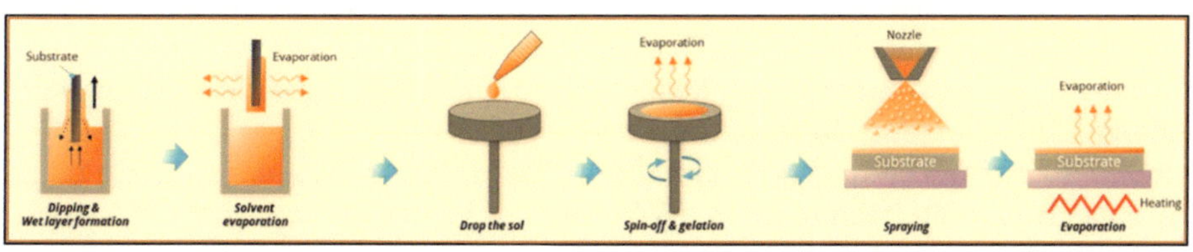

Figure 4.
Sol-gel method.

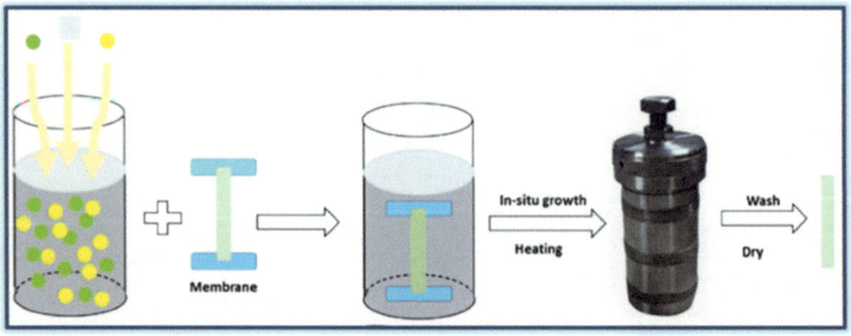

Figure 5.
Hydrothermal method.

quantum dots [11]. In a typical process, a precursor solution containing cadmium and selenium salts is sealed in the autoclave and heated under controlled conditions. The elevated temperature and pressure promote the growth of crystalline CdSe nanoparticles. Hydrothermal synthesis allows for precise control over the size and properties of the resulting chalcogenide nanomaterials, making it essential in nanotechnology and the development of quantum dot-based technologies, including photodetectors, solar cells, and biological imaging agents.

2.2 Microscale synthesis

Microscale synthesis of chalcogenides involves the creation of chalcogenide materials on a miniature to small scale. One of the primary methods used for this purpose is physical vapour deposition (PVD), where chalcogenide materials are produced through processes like evaporation and sputtering. In PVD, chalcogenide materials are vaporized and then condensed onto a substrate, forming thin films or coatings with precise thickness and high uniformity. This approach is crucial for applications like integrated circuits and optical devices, where microscale chalcogenide materials are utilized to develop advanced electronics, telecommunications components, and optical coatings, taking advantage of their unique electrical and optical properties.

2.2.1 Physical vapor deposition (PVD)

Physical vapor deposition (PVD) is a widely used technique for depositing chalcogenide thin films onto substrates. The PVD process is shown in **Figure 6**. One common example of PVD in chalcogenide technology is the fabrication of phase-change memory devices [12]. In this process, a chalcogenide material, often a blend of elements like germanium, antimony, and tellurium (GST), is heated to create vapour in a vacuum chamber. The vapour then condenses on a substrate, forming a thin chalcogenide film. This film's properties can be controlled by adjusting deposition parameters like temperature and pressure. Phase-change memory devices exploit the unique properties of chalcogenides that can rapidly switch between amorphous and crystalline phases, making them invaluable in non-volatile memory technologies. PVD ensures the precise deposition of chalcogenide materials, enabling the development of advanced memory storage devices in the electronics industry.

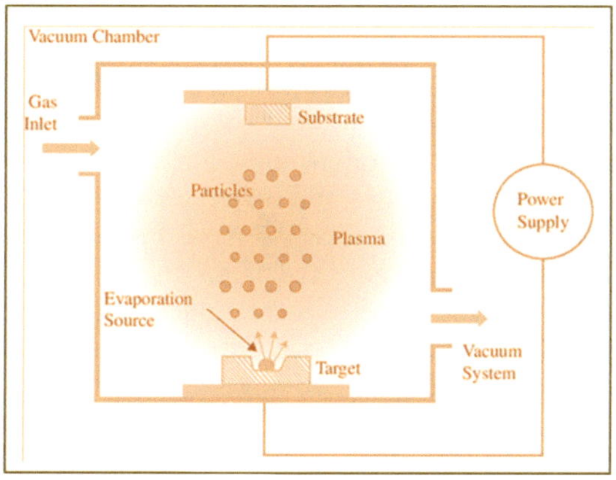

Figure 6.
Physical vapor deposition.

2.2.2 Chemical vapor deposition (CVD)

Chemical vapor deposition (CVD) is a versatile technique for the synthesis of chalcogenide thin films and coatings. **Figure 7** shows the working mechanism of CVD method. In CVD, chalcogenide precursor gases are thermally decomposed on a substrate, forming chalcogenide materials in a controlled manner. This method offers precise control over film thickness, composition, and microstructure, making it invaluable for applications like semiconductor device manufacturing, optical devices, and surface coatings [13]. By adjusting parameters such as temperature, pressure, and precursor gases, CVD enables the engineering of chalcogenide materials with tailored properties, playing a crucial role in various industries where high-quality thin films are essential for advanced electronics, optics, and protective coatings.

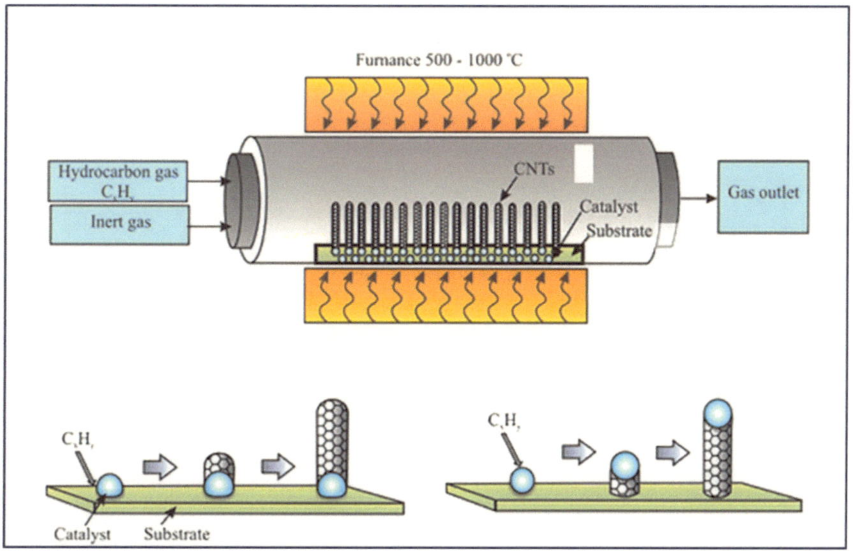

Figure 7.
Chemical vapor deposition.

2.3 Macroscale synthesis

Macroscale synthesis of chalcogenides involves the production of bulk chalcogenide materials for a wide range of applications. Techniques such as solid-state synthesis and chemical reactions are typically employed to create chalcogenides on a larger scale. In solid-state synthesis, chalcogen and metal precursors are mixed and heated to high temperatures, enabling chemical reactions that result in the formation of bulk chalcogenide compounds. This method is crucial in the production of materials for applications like thermoelectric devices, superconductors, and infrared optics, where macroscopic chalcogenide materials with specific properties are needed. Macroscale chalcogenide synthesis contributes to the advancement of various industries, such as energy, electronics, and defense, where these materials play a pivotal role in enabling innovative technologies and products.

2.3.1 Solid-state synthesis

Solid-state synthesis is a fundamental method for the bulk production of chalcogenides as shown in **Figure 8**. In this process, chalcogen and metal precursors are mixed and heated to high temperatures, causing chemical reactions that lead to the formation of chalcogenide compounds [14]. This technique is widely used to create bulk chalcogenide materials for various applications, such as thermoelectric devices, superconductors, and infrared optics. Solid-state synthesis allows for the controlled fabrication of macroscopic chalcogenide materials with tailored properties, enabling the development of advanced materials for industries like energy, electronics, and defense.

2.3.2 Zone refining

Zone refining is a specialized purification technique applied to high-purity macroscale chalcogenide materials. In this method, a molten zone is gradually moved along a solid rod or ingot of chalcogenide material, selectively segregating impurities from the crystalline structure [15]. As the impurities are pushed to the end of the rod, a high-purity region is left behind. Zone refining is particularly useful in industries where ultrapure chalcogenides are essential, such as the semiconductor industry,

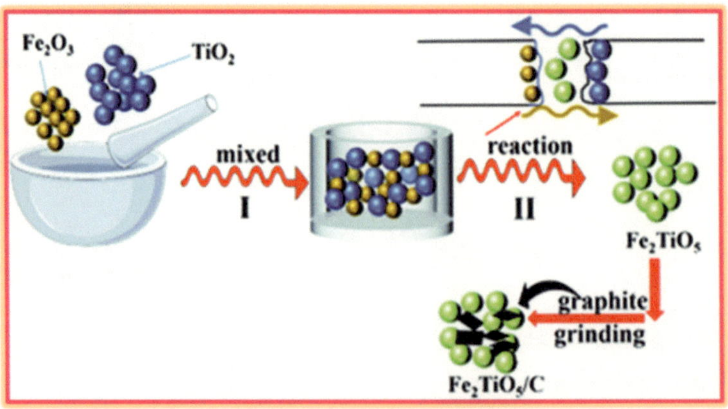

Figure 8.
Solid state synthesis.

where even trace impurities can significantly impact the performance of electronic devices. This method ensures the production of high-purity chalcogenide materials that meet stringent quality requirements, making it a crucial process for the manufacturing of advanced electronic components and other applications demanding exceptional material purity.

2.3.3 Crystal growth

Crystal growth of chalcogenides is a vital process in producing high-quality single crystals or crystalline materials with specific properties [16]. Techniques such as Bridgman-Stockbarger, Czochralski, and the floating zone method are commonly employed. The different crystal growth methods are shown in **Figure 9**. These methods involve the controlled cooling of chalcogenide melts or the growth of crystals from a seed crystal in a controlled atmosphere. Crystal growth allows to produce well-ordered chalcogenide structures, which are essential for applications in solid-state electronics, lasers, and photodetectors, where the precise arrangement of atoms in the crystal lattice influences the material's electrical, optical, and thermal properties.

2.3.4 Melt quenching

Melt quenching is a fundamental method for producing chalcogenide glasses, amorphous materials that incorporate chalcogen elements (sulfur, selenium, tellurium) and metals or metalloids. The whole process is shown in **Figure 10**. In this process, a controlled blend of these constituents is heated to high temperatures until it becomes a molten liquid. The molten chalcogenide mixture is then rapidly cooled or "quenched" to room temperature, preventing the formation of a crystalline structure and locking the material into an amorphous state [17]. Notable examples include the synthesis of arsenic trisulphide (As_2S_3) chalcogenide glass, which is widely used in infrared optics and optical fibers due to its excellent transparency in the infrared region. Another example is the production of germanium selenide (GeSe) glasses, which are essential in non-linear optics and photonic devices. Melt quenching offers

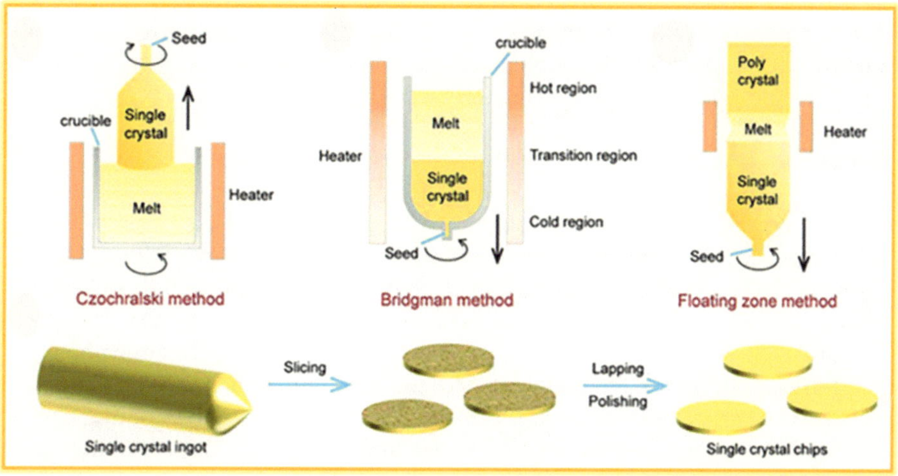

Figure 9.
Crystal growth methods.

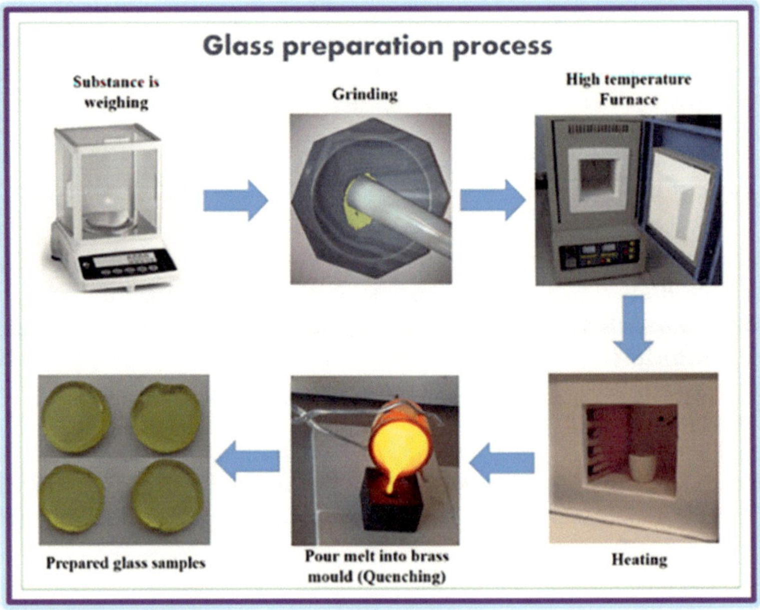

Figure 10.
Melt quenching.

precise control over chalcogenide glass composition, enabling the tailoring of optical and electrical properties for various applications, ranging from thermal imaging to data transmission systems.

3. Applications based on magnitudes of the material

Chalcogenides find applications across various scales, from the nanoscale to the macroscale, offering a wide range of possibilities in different fields. Chalcogenides offer a versatile toolkit for scientists and engineers to tailor materials to meet specific requirements, whether at the nanoscale for cutting-edge nanotechnology or at the macroscale for practical applications that impact our daily lives. Their versatility is reflected in their wide-ranging applications across different scales.

3.1 Nanoscale

- *Nanoelectronics*: Chalcogenides play a pivotal role in nanoelectronics, where their unique properties are harnessed for a wide range of applications [18]. One prominent example is the utilization of phase-change chalcogenide materials like GeSbTe (germanium-antimony-tellurium) in rewritable optical and electrical storage devices. These materials can rapidly switch between amorphous and crystalline states, enabling data storage in products such as DVDs and phase-change memory devices. Chalcogenide-based resistive switching devices, exemplified by Ag-Ge-S or Cu-Te-S based compounds, are employed in non-volatile memory technologies like resistive random-access memory (ReRAM). Chalcogenides also find use in emerging field-effect transistors, as in the case of black phosphorus-chalcogenide heterostructures, for their superior charge transport properties. These examples underscore the significant impact of chalcogenides on the development of cutting-edge nano electronic components and memory technologies.

- *Quantum dots*: Quantum dots, which are nanoscale semiconductor structures, often employ chalcogenides and are influential in various applications [19]. Chalcogenide quantum dots, such as cadmium selenide (CdSe) or lead selenide (PbSe), exhibit unique electronic and optical properties due to quantum confinement effects. They are used extensively in nanotechnology and biological imaging. For instance, CdSe quantum dots emit bright and tunable fluorescence, making them valuable in fluorescent tagging and tracking of biological molecules and cells, and in quantum dot solar cells, where they efficiently convert sunlight into electricity. PbSe quantum dots, on the other hand, are well-suited for infrared photodetectors and sensors due to their sensitivity in the near-infrared region. These examples emphasize the critical role of chalcogenide quantum dots in advancing fields like nanotechnology, biotechnology, and renewable energy.

- *Nanowires*: Chalcogenide nanowires are one of the most promising nanostructures, with applications ranging from electronics to sensing [20]. For instance, tellurium nanowires, produced through methods like the vapor-liquid-solid (VLS) technique, are used in phase-change memory devices, offering high-speed data storage and low power consumption. In thermoelectric applications, bismuth chalcogenide nanowires, such as Bi2Te3 and Sb2Te3, exhibit excellent thermoelectric properties, enabling efficient energy conversion from heat differentials. Additionally, chalcogenide nanowires have proven valuable in sensor technologies, such as zinc selenide nanowires used in highly sensitive gas sensors. These examples showcase the versatility and functionality of chalcogenide nanowires, which are central to many emerging technologies and hold great promise in enhancing the performance of various devices.

- *Catalysis*: Chalcogenides play a significant role in catalysis, particularly in hydrodesulphurization (HDS) of crude oil. Transition metal chalcogenides, like molybdenum disulfide (MoS_2) and tungsten disulfide (WS_2), are essential components in HDS catalysts that remove sulfur impurities from petroleum products, improving their environmental and safety qualities. These catalysts promote the desulphurization process by breaking the sulfur-carbon bonds in hydrocarbons [21]. Moreover, chalcogenide materials have also demonstrated promise in electrocatalysis, where they are used in hydrogen evolution reactions, contributing to the development of clean and sustainable energy technologies. These examples underscore the critical role of chalcogenides in various catalytic processes, including those essential to produce cleaner fuels and the advancement of renewable energy sources.

3.2 Microscale

- *Integrated circuits*: Chalcogenides are increasingly finding application in integrated circuits (ICs), where their unique electrical properties are harnessed for memory and logic devices [22]. Phase-change chalcogenides like GeSbTe are a prime example, as they form the basis of non-volatile phase-change memory (PCM) devices. These PCM devices offer advantages like fast read/write speeds, low power consumption, and high endurance, making them integral in ICs, including storage-class memory in data centres. Chalcogenide-based selector devices, such as the ovonic threshold switch (OTS), are also crucial for building 3D NAND flash memory arrays, where their non-linear current-voltage characteristics are instrumental in memory cell isolation and addressing. The integration of chalcogenides into ICs represents a significant advancement in electronics, enabling more efficient, compact, and higher-capacity memory and logic devices for a wide range of applications, from smartphones to data storage solutions.

- *Optical devices*: Chalcogenides play a fundamental role in the development of advanced optical devices due to their exceptional optical properties. Chalcogenide glass fibers and waveguides are crucial components in optical communication systems, allowing the efficient transmission of signals in the infrared region [23]. Chalcogenide materials also serve as the basis for infrared lenses and windows, essential in thermal imaging cameras and sensors. Additionally, chalcogenide micro structured optical fibers, like photonic crystal fibers, are used for supercontinuum generation, enabling broad-spectrum light sources for applications in spectroscopy, medical imaging, and sensing. Their superior nonlinear optical properties make chalcogenides indispensable in the creation of devices like optical parametric oscillators, which produce tunable and coherent light for a variety of scientific and industrial applications. These examples highlight the central role of chalcogenides in advancing optical technologies and applications.

- *Infrared optics*: Chalcogenides are integral to the field of infrared optics, offering unparalleled capabilities for capturing and manipulating thermal radiation in various applications. Chalcogenide materials, such as amorphous chalcogenide glasses, are used to create lenses and windows that efficiently transmit infrared radiation while maintaining high optical quality [24]. These components are essential in thermal imaging cameras for applications like surveillance, defense, and industrial inspection. Moreover, chalcogenide materials like zinc selenide (ZnSe) are used in the construction of optical components for carbon dioxide (CO_2) laser systems, which are employed in materials processing, surgery, and atmospheric monitoring. The exceptional transmittance of chalcogenides in the infrared region allows for precise, high-resolution imaging and sensing, making them indispensable in the development of state-of-the-art infrared optics.

3.3 Macroscale

- *Photovoltaics*: Chalcogenides have become prominent players in the field of photovoltaics, particularly in the development of thin-film solar cells. Notable examples include cadmium telluride (CdTe) and copper indium gallium selenide (CIGS), which are used as the active absorber materials in these solar cells. CdTe-based solar panels have seen widespread adoption due to their cost-effectiveness and efficiency in converting sunlight into electricity [25]. CIGS solar cells offer a versatile alternative, with tunable composition for improved performance. The use of chalcogenides in photovoltaics underscores their pivotal role in renewable energy technology, contributing to the transition to sustainable power sources and reducing our reliance on fossil fuels.

- *Semiconductor devices*: Certainly! Chalcogenides are indispensable in semiconductor devices, serving as the foundation for various components crucial to modern electronics. Phase-change chalcogenide materials, such as GeSbTe, are central to rewritable optical and electrical storage devices, including CDs, DVDs, and phase-change memory [26]. These materials can rapidly switch between amorphous and crystalline states, allowing for high-speed data storage and retrieval. Chalcogenide-based non-volatile memory technologies, such as resistive random-access memory (ReRAM) with Ag-Ge-S or Cu-Te compounds, are gaining attention for their potential in future electronic applications. The unique electrical and optical properties of chalcogenides are also leveraged in photodetectors, infrared sensors, optical components for night vision technology, telecommunications, and high-speed data communication. CdTe-based radiation detectors, on the other hand, capitalize on cadmium telluride's high sensitivity to ionizing radiation, making them pivotal for applications in nuclear science, medical imaging, homeland security, and industrial radiography. Chalcogenides play a foundational role in semiconductor devices, contributing to a wide range of electronic, photonic, and radiographic applications.

- *Thermoelectric materials*: Chalcogenides have emerged as crucial materials in the field of thermoelectric power generation and solid-state cooling. Certain chalcogenide compounds, such as bismuth telluride (Bi_2Te_3) and lead telluride (PbTe), exhibit excellent thermoelectric properties [27]. These materials can efficiently convert heat differentials into electricity, making them vital in thermoelectric generators for applications like waste heat recovery and remote power sources. In addition, chalcogenides have found use in thermoelectric coolers, which rely on the Peltier effect to provide compact, solid-state cooling solutions. For instance, bismuth antimony telluride (BiSbTe) compounds are employed in portable refrigeration and microelectronics thermal management. These examples highlight the significance of chalcogenides in enhancing energy efficiency and offering sustainable cooling solutions in various industries.

- *Superconductors*: Chalcogenides have played a significant role in the development of high-temperature superconductors (HTS), a ground-breaking class of materials that can conduct electricity without resistance at elevated temperatures [28]. Yttrium barium copper oxide ($YBa_2Cu_3O_7$) and bismuth strontium calcium copper oxide (BSCCO) are two prominent examples of chalcogenide based HTS materials. These compounds exhibit superconducting properties at temperatures much higher than traditional low-temperature superconductors, making them attractive for practical applications. Chalcogenide-based HTS materials are employed in a range of technologies, including the development of more efficient power transmission lines and magnetic resonance imaging (MRI) machines, where superconducting magnets provide high magnetic fields for medical diagnostics. These advances in HTS, enabled by chalcogenides, have the potential to revolutionize multiple industries by improving energy efficiency and enabling novel technologies.

- *Infrared imaging*: Chalcogenides are integral to the field of infrared imaging, providing optical materials crucial for capturing and visualizing thermal radiation. Chalcogenide glass, like germanium selenide (GeSe) and zinc sulphide (ZnS), is commonly used in the construction of lenses and windows for thermal imaging cameras and sensors [29]. These components allow the transmission of infrared radiation while maintaining high optical quality, facilitating the detection of temperature variations in objects and environments. The versatility of chalcogenide materials is evident in their use in applications such as surveillance, defense, and industrial inspection, where thermal imaging technology enhances situational awareness, safety, and predictive maintenance, exemplifying their pivotal role in advancing infrared imaging capabilities.

- *Lubricants and coatings*: Chalcogenides have found applications in lubricants and coatings, primarily due to their exceptional wear-resistant properties. Molybdenum disulfide (MoS_2), a common chalcogenide compound, is often used as a solid lubricant in extreme conditions, where its layered structure reduces friction and minimizes wear in components like automotive engine parts and industrial machinery [30]. Additionally, chalcogenide coatings, such as diamond-like carbon (DLC) films containing chalcogenide elements, provide excellent anti-wear and anti-corrosion properties. These coatings are used on various surfaces, including cutting tools, aerospace components, and medical devices, to enhance their durability and reduce friction. Chalcogenides' role in lubrication and coating technologies underscores their significance in prolonging the lifespan and performance of a wide range of mechanical systems and devices.

- *Chalcogenide glasses*: Chalcogenide glasses are a fascinating class of materials composed primarily of chalcogens like sulfur, selenium, or tellurium, often in combination with other elements. These glasses exhibit unique optical, electrical,

Scale of material	Synthesis methods	Applications	Examples
Nano	• Chemical precipitation • Sol-gel method • Hydrothermal synthesis	Nanoelectronics	GeSbTe AgGeS CuTeS
		Quantum Dots	PbSe CdSe
		Nanowires	Bi2Te3 Sb2Te3
		Catalysis	MoS_2 WS_2
Micro	• Physical Vapor Deposition (PVD) • Chemical Vapor Deposition (CVD)	Integrated Circuits	GeSbTe
		Optical Devices	ZnTe
		Infrared Optics	ZnSe
Macro	• Solid-state synthesis • Zone refining • Crystal Growth • Melt quenching	Superconductors	$YBa_2Cu_3O_7$ BSCCO
		Infrared Imaging	GeSe ZnS
		Lubricants and Coatings	MoS_2
		Chalcogenide glasses	As_2S_3 GeSbTe

Table 1.
Comparison of chalcogenides and its various forms.

and thermal properties, making them integral to various technological applications [31]. An exemplary chalcogenide glass is amorphous arsenic trisulphide (As_2S_3), widely used in optical components such as optical fibers, lenses, and windows for infrared applications. Additionally, chalcogenide glasses are pivotal in the development of phase-change memory devices, where materials like germanium-antimony-tellurium (GST) serve as the active medium, switching between amorphous and crystalline states to store data. Their ability to efficiently transmit infrared radiation, combined with their non-crystalline structure, makes chalcogenide glasses indispensable in optical communication, data storage, and thermal imaging, emphasizing their versatile role in modern technology (**Table 1**).

4. Conclusion

In conclusion chalcogenides hold significant societal applications across various forms, contributing to advancements in technology, energy, and healthcare. In the field of electronics, chalcogenide materials are crucial for non-volatile memory devices, enhancing the efficiency and storage capacity of electronic devices. In the renewable energy sector, chalcogenides, particularly in the form of thin-film solar cells like cadmium telluride (CdTe) and copper indium gallium selenide (CIGS), play a pivotal role in harnessing solar energy for sustainable power generation. Chalcogenide glasses, known for their infrared transparency, find applications in optical fibers for high-speed telecommunications, facilitating global connectivity. Furthermore, the development of chalcogenide-based sensors and detectors contributes to advancements in healthcare, environmental monitoring, and security systems. The societal impact of chalcogenides is underscored by their diverse applications, improving everyday life through technological innovations, sustainable energy solutions, and enhanced communication systems. Our Present work will give an overview of the importance of different-scale synthesis of chalcogenides lies in its capacity to precisely tailor material properties, enabling diverse applications from nanoscale electronics to macroscale energy solutions.

Author details

Manivel Rajan[1], Raja Arumugam[2*], Sivasubramani Vediyappan[3], Siva Vadivel[4], Rajesh Paulraj[1] and Ramasamy Perumalsamy[1]

1 Department of Physics, Sri Sivasubramaniya Nadar College of Engineering, Chennai, Tamil Nadu, India

2 CNR-SPIN, C/O University of Salerno, Fisciano, Salerno, Italy

3 Department of Physics, School of Engineering and Technology, Dhanalakshmi Srinivasan University, Tiruchirapalli, Tamil Nadu, India

4 Department of Physics, Karpagam Academy of Higher Education, Coimbatore, Tamil Nadu, India

*Address all correspondence to: rajaphy014@gmail.com

References

[1] Xiao JR, Yang SH, Feng F, Xue HG, Guo SP. A review of the structural chemistry and physical properties of metal chalcogenide halides. Coordination Chemistry Reviews. 2017;**347**:23-47. DOI: 10.1016/j.ccr.2017.06.010

[2] Guziec FS, Sanfilippo LJ. Tetrahedron Report Number 242 Synthetically Useful Extrusion Reactions of Organic Sulphur, Selenium and Tellurium Compounds. Amsterdam, Netherlands: Elsiever; 1988

[3] Cao T, Cen M. Fundamentals and applications of chalcogenide phase-change material photonics. Advanced Theory and Simulations. Wiley-VCH Verlag. 01 Aug 2019;**2**(8):1-17. DOI: 10.1002/adts.201900094

[4] Freitas JN, Gonçalves AS, Nogueira AF. A comprehensive review of the application of chalcogenide nanoparticles in polymer solar cells. Nanoscale. 2014;**6**(12):6371-6397. DOI: 10.1039/c4nr00868e

[5] Lee TH, Elliott SR. Chemical bonding in chalcogenides: The concept of multicenter Hyperbonding. Advanced Materials. 2020;**32**(28):1-9. DOI: 10.1002/adma.202000340

[6] Zakery A, Elliott SR. Optical properties and applications of chalcogenide glasses: A review. Journal of Non-Crystalline Solids. 2003;**330**(1-3):1-12. DOI: 10.1016/j.jnoncrysol.2003.08.064

[7] Beun JA, Nitsche R, Lichtensteiger M. Optical and Electrical Properties of Ternary Chalcogenides. Vol. 27. Issue 5. Netherlands: Elsevier; May 1961. pp 448-452. DOI: 10.1016/0031-8914(61)90002-7

[8] Ding J, Liu C, Xi L, Xi J, Yang J. Thermoelectric transport properties in chalcogenides ZnX (X=S, Se): From the role of electron-phonon couplings. Journal of Materiomics. 2021;7(2):310-319. DOI: 10.1016/j.jmat.2020.10.007

[9] Yadav RS, Pandey AC, Sanjay SS. Optical Properties of Europium Doped Bunches of Zno Nanowires Synthesized by Co-Precipitation Method. Romania: Virtual Company of Physics SRL; 2009

[10] Tohge N, Asuka M, Minami T. Non-Crystalline Solids sol-Gel Preparation and Optical Properties of Silica Glasses Containing Cd and Zn Chalcogenide Microcrystals. Vol. 147-148. Netherlands: Elsevier; 1992. pp. 652-656

[11] Rajamathi M, Seshadri R. Oxide and chalcogenide nanoparticles from hydrothermal/solvothermal reactions. In: Current Opinion in Solid State and Materials Science. Vol. 6. Netherlands: Elsevier; 2002. pp. 338-345

[12] Andzane J et al. Thickness-dependent properties of ultrathin bismuth and antimony chalcogenide films formed by physical vapor deposition and their application in thermoelectric generators. Materials Today Energy. Mar 2021;**19**:1-37. DOI: 10.1016/j.mtener.2020.100587

[13] Xie C, Yang P, Huan Y, Cui F, Zhang Y. Roles of salts in the chemical vapor deposition synthesis of two-dimensional transition metal chalcogenides. Dalton Transactions. 2020;**49**(30):10319-10327. DOI: 10.1039/d0dt01561j

[14] Vaidhyanathan B, Ganguli M, Rae K. Fast Solid State Synthesis of Metal Vanadates and Chalcogenides Using Microwave Irradiation. Vol. 30. no. 9. Netherlands:

Elsevier; 1995. pp. 1173-1177. DOI: 10.1016/0025-5408(95)00099-2

[15] Maier H, Daniel DR, Herkert R. Liquid Encapsulation Zone-Refining of PbS. Berlin, Germany: Springer Nature; 1978

[16] Triboulet R. 5 Growth of Zinc Chalcogenides 15.1 Introduction. Netherlands: Elsevier; 2003. pp. 497-523. DOI: 10.1016/B978-081551453-4.50017-8

[17] Abd-Elnaiem AM, Abbady G. A thermal analysis study of melt-quenched Zn5Se95 chalcogenide glass. Journal of Alloys and Compounds. Mar 2020;**818**:1-34. DOI: 10.1016/j.jallcom.2019.152880

[18] Kim Y et al. Atomic-layer-deposition-based 2D transition metal chalcogenides: Synthesis, modulation, and applications. Advanced Materials. John Wiley and Sons Inc.; 01 Nov 2021;**33**(47):1-33. DOI: 10.1002/adma.202005907

[19] Gui R, Jin H, Wang Z, Tan L. Recent advances in synthetic methods and applications of colloidal silver chalcogenide quantum dots. Coordination Chemistry Reviews. Elsevier. 2015;**296**:91-124. DOI: 10.1016/j.ccr.2015.03.023

[20] Huang L, Lu JG. Synthesis, characterizations and applications of cadmium chalcogenide nanowires: A review. Journal of Materials Science and Technology. Jun 2015;**31**(6):556-572. DOI: 10.1016/j.jmst.2014.12.005

[21] Shao X et al. Metal chalcogenide-associated catalysts enabling CO2electroreduction to produce low-carbon fuels for energy storage and emission reduction: Catalyst structure, morphology, performance, and mechanism. Journal of Materials Chemistry A. 2021;**9**(5):2526-2559. DOI: 10.1039/d0ta09232k

[22] Czubatyj W, Hudgens SJ. Invited paper: Thin-film ovonic threshold switch: Its operation and application in modern integrated circuits. Electronic Materials Letters. 2012;**8**(2):157-167. DOI: 10.1007/s13391-012-2040-z

[23] Eggleton BJ, Luther-Davies B, Richardson K. Chalcogenide photonics. Nature Photonics. 2011;**5**(3):141-148. DOI: 10.1038/nphoton.2011.309

[24] Adam JL, Calvez L, Trolès J, Nazabal V. Chalcogenide glasses for infrared photonics. International Journal of Applied Glass Science. 2015;**6**(3):287-294. DOI: 10.1111/ijag.12136

[25] Saha S, Johnson M, Altayaran F, Wang Y, Wang D, Zhang Q. Electrodeposition fabrication of chalcogenide thin films for photovoltaic applications. Electrochem. 2020;**1**(3):286-321. DOI: 10.3390/electrochem1030019

[26] Mehta N. Applications of Chalcogenide Glasses in Electronics and Optoelectronics: A Review. India: CSIR-NIScPR; 2006

[27] Meng H, An M, Luo T, Yang N. Thermoelectric applications of chalcogenides. In: Chalcogenide: From 3D to 2D and beyond. Netherlands: Elsevier; 2019. pp. 31-56. DOI: 10.1016/B978-0-08-102687-8.00002-6

[28] Tsendin KD, Denisov DV. Application of Chalcogenides for Creation of New Superconductors. Romania: National Institute of Optoelectronics; 2003

[29] Cha DH et al. Effect of Temperature on the Molding of Chalcogenide Glass Lenses for Infrared Imaging Applications. Washington, USA: Optica Publishing Group; 2010

[30] Yang JF, Parakash B, Hardell J, Fang QF. Tribological properties of

transition metal di-chalcogenide based lubricant coatings. Frontiers of Materials Science. 2012;**6**(2):116-127. DOI: 10.1007/s11706-012-0155-7

[31] Kolobov A, Tominaga J, Kolobov AV, Tominaga J. Chalcogenide Glasses in Optical Recording: Recent Progress. Romania: National Institute of Optoelectronics; 2002. Available from: https://www.researchgate.net/publication/266161075